Cerebral Circulation and Stroke

Editor: K. J. Zülch

With Contributions by

A. Agnoli, Rome · L. Arbus, Toulouse · W. Bartsch, Munich · A. Bes, Toulouse · J. Espagno, Toulouse · C. Fazio, Rome · W. S. Fields, Houston · C. Fieschi, Siena · H. Gastaut, Marseille · J. C. Gautier, Paris · K.-A. Hossmann, Cologne · V. Hossmann, Cologne · D. H. Ingvar, Lund · S. Katsuky, Fukuoka · S. S. Kety, Boston · K. Kimoto, Fukuoka · P. Kleihues, Cologne · M. Kribs, Cologne · G. Lazorthes, Toulouse · Y. Lazorthes, Toulouse · F. Lhermitte, Paris · C. Loeb, Genoa · J. Marshall, London · H. Metzinger, Cologne · J. S. Meyer, Houston · R. Naquet, Marseille · K. Nishimaru, Fukuoka-Shi · Y. Nishino, Fukuoka-Ken · N. Okabe, Fukuoka · Y. Olsson, Gothenburg · T. Omae, Fukuoka-Shi · A. Pierach, Bad Nauheim · R. W. Ross Russell, London · G. Salamon, Marseille · R. L. Swank, Portland · M. Toga, Marseille · M. Toyoda, Tokyo · R. A. Vigouroux, Marseille · E. J. Wylie, San Francisco · T. Yamaguchi, Fukuoka-Shi · K. J. Zülch, Cologne

With 71 Figures

Springer-Verlag Berlin · Heidelberg · New York 1971

KLAUS JOACHIM ZÜLCH, Professor of Neurology and Director of the Division of General Neurology at Max-Planck-Institute for Brain Research and the Department of Neurology, City Hospital, D-5000 Köln-Merheim

ISBN-13: 978-3-642-95176-3   e-ISBN-13: 978-3-642-95174-9
DOI: 10.1007/978-3-642-95174-9

Preface

Cerebrovascular disease is today the cause of death in a fifth of the population in countries of modern technology. It seems legitimate that society expects medicine to analyse the causes of this menace and, as far as it is possible, to cure its sequelae and prevent its origin. Even when the patients survive, sequelae such as hemi-or monoplegia, aphasia and visual deficits severely limit the mode of life of those afflicted.

This book tries to show the methods by which medicine is striving to investigate and analyse the causes of this important condition—cerebrovascular disease.

For helping us in our efforts to overcome cerebrovascular disease, I wish to thank all those who contributed to this most important problem. I am most grateful to Drs. G. R. JOHNSON/Columbus (Ohio), B. MAIR/London and J. H. MILHORAT/Philadelphia for their help in translation and to the publisher, Springer, for the excellent lay-out of this book.

Cologne January, 1971   K. J. ZÜLCH

Table of Contents

Clinical Features and Pathogenesis of Cerebrovascular Disease: Hypertension and Hypotension

Therapy

Author Index

KLEIHUES, P., Dr., Max-Planck-Institut für Hirnforschung, Abt. für Allgemeine Neurologie, D-5000 Köln 91, Ostmerheimer Str. 200

KRIBS, M., Dr., Max-Planck-Institut für Hirnforschung, Abt. für Allgemeine Neurologie, D-5000 Köln 91, Ostmerheimer Str. 200

LAZORTHES, G., Prof. Dr., Centre Hospitalier et Universitaire de Toulouse, Hôpital de Purpan, Clinique de Neurochirurgie, F-31 Toulouse

LAZORTHES, Y., Dr., Chef de Clinique, Assistant des Hôpitaux, C.H.U. Purpan, F-31 Toulouse

LHERMITTE, F., Prof. Dr., Hôpital de la Salpêtrière, Clinique des Maladies du Système Nerveux, 47, Bd. de L'Hôpital, F-75 Paris XIII

LOEB, C., Prof. Dr., Clinica delle Malattie Nervose e Mentali dell'Università di Genova, Via A. De Toni 5, I-16132 Genova

MARSHALL, J., Dr., Institute of Neurology, The National Hospital, Queen Square, London W.C. 1/England

METZINGER, H., Dr., Max-Planck-Institut für Hirnforschung, Abt. für Allgemeine Neurologie, D-5000 Köln 91, Ostmerheimer Str. 200

MEYER, J. S., Prof. Dr., Baylor University, College of Medicine, Depart. of Neurology, Houston, TX 77021/USA

NAQUET, R., Dr., Maitre de Recherches au C.N.R.S., I.N.P. 3, 31 chemin Joseph-Aiguier, F-13 Marseille IX

NISHIMARU, K., M.D., Staff Second Department of Internal Medicine, Faculty of Medicine, Kyushu University, 1276, Katakasu, Fukuoka-Shi/Japan

NISHINO, Y., Assistant Staff Physician, Sasaguri Hospital, Sasaguri-Machi, Kasuya-Gun, Fukuoka-Ken/Japan

OKABE, N., Prof. of Medicine, Kyushu Dental College, Kita-Kyushu City/Japan, and Associate Clinical Prof. of Medicine, Faculty of Medicine, Kyushu University, Fukuoka City/Japan

OLSSON, Y., Prof. Dr., Göteborgs Universitet, Patologiska Institutionen Sahlgrenska Sjukhuset 41345 Göteborg/Schweden

OMAE, T., M.D., Senior Instructor, Kyushu University Hospital, 1276, Katakasu, Fukuoka-Shi/Japan

PIERACH, A., Prof. Dr., D-6350 Bad Nauheim, Königsberger Str. 5

ROSS RUSSELL, R. W., Prof. Dr., St. Thomas' Hospital, Dept. of Neurology, London S.E.1/England

SALAMON, G., Prof. à la Faculté de Médecine, Service de Neuroradiologie, Hôpital de La Timone, Bd. Jean-Moulin, F-13 Marseille V

SWANK, R. L., Dr., University of Oregon, Medical School, Division of Neurology, 3181 S.W. Sam Jackson Park Road, Portland, Oregon 97201/USA

TOGA, M., Dr., Prof. à la Faculté de Médecine Hôpital de la Timone, F-13 Marseille V

TOYODA, M., Dr., Research Associate and Instructor, Department of Neurology, Keio University, Tokyo/Japan

VIGOUROUX, R. A., Dr., Assistant des Hôpitaux, Hôpital de la Timone, Service de Neurologie, F-13 Marseille V

WYLIE, E. J., Prof. Dr., University of California, San Francisco Medical Center, Dept. of Surgery, San Francisco, CA 94122/USA

YAMAGUCHI, T., Assistant Staff, Second Department of Internal Medicine, Faculty of Medicine, Kyushu University, 1276, Katakasu, Fukuoka-Shi/Japan

ZÜLCH, K. J., Prof. Dr., Direktor des Max-Planck-Institutes für Hirnforschung, Abt. Allgemeine Neurologie und der Neurologischen Klinik der Städtischen Krankenanstalt Köln-Merheim, D-5000 Köln 91, Ostmerheimer Str. 200

Introduction

K. J. Zülch

This brief monograph aims to show present trends in research and clinical work in cerebrovascular disease. It is not meant to be a complete review of the present knowledge, which would require an extensive textbook. The pace of progress in the last decade has been so rapid that it is necessary to reconsider and observe the work of others. This has been attempted in many yearly or biyearly national or international conferences, but several years may elapse before the proceedings are published and even then only the participants receive them or the text may be too special.

The need exists for a small volume to show what is going on in the field of cerebrovascular disease. Practically all the authors approached have submitted short articles, either summarizing their own work, or describing a particular problem which shows the line of research being followed. The various essays may seem somewhat unrelated at first sight, but we considered that a more colourful book would result from allowing the contributors to follow their own approach in their papers. The following introductory chapter is designed to correlate them, where gaps exist. Moreover, some personal opinions and experiences from the work of our own team are added.

The up to date discussion in the field of cerebral circulation and stroke revolves around the pathogenesis of the regional "cerebrovascular insufficiency", i.e. the local disturbance of blood flow. This assumes some knowledge of the normal and disturbed circulation and the typical haemodynamical patterns of cerebrovascular insufficiency. Regional predilections such as the "endangered territories" (terminal and/or frontier zones and "last fields") have to be considered. The causes for such local disturbances—which may be "local" stenosing factors or "general" causes detrimental to the heart's action and peripheral circulation—have to be considered. The "living morphology" of the neuro-radiological pictures helps to visualise the changes already taking place in life and particular methods such as electroencephalography and scanning aid in analysing the functional state of the tissue in regional arterial insufficiency. Changes of viscosity and the metabolic efficiency of the blood may augment the hypoxic state, to name only some essential features of the very complex features of such a disturbance, which are only superficially expressed by the haemodynamics or the changes in blood pressure.

The three main theories for the regional cerebrovascular insufficiency are very controversial and each may be mainly responsible for one particular type of stroke. Integrated studies from as many aspects as possible are essential to clarify the unsolved problem of "strokes".

The editor has not hesitated to present very divergent interpretations of pathogenesis of cerebrovascular insufficiency in the hope that a broader discussion will help to clarify the pathogenesis and promote a useful programme for therapy and, if possible, the prevention of "strokes" and their minor antecedents.

In this introduction we commence with a definition of cerebrovascular insufficiency and a review of the leading concepts of its pathogenesis.

"Cerebrovascular insufficiency" was introduced by Corday and Rothenberg (1957) following Büchner's (1939) and Rein's (1931) concept in cardiac circulation and coronary infarct, where they used the term "coronary insufficiency". It presupposes then a general or local defect of the circulation which becomes thereby insufficient to cope with either the "functional" or the "morphological" metabolism (Opitz and Schneider, 1950) of the organ, in our case the brain. The term insufficiency may define a transient or chronic and permanent disturbance. It does not presuppose any time factor. The causes may be predominantly "general"—a deficiency of general circulation—or "local" i.e. regional. The local factors may be a) occlusion or b) stenosis of a vessel lumen and possibly deficiency of an existing "collateral" circulation in a) or b).

What types of "anastomoses" and collateral circulation, which actually function, occur in the brain? Cohnheim's (1872) concept of the cerebral "end-arteries" has dominated for long but was already rejected by Heubner in 1874 and had to be rediscovered in 1953 by van der Eecken and Adams, as it never gained its correct position as a pathogenetic concept. Now we know the significance of the "meningeal anastomoses" of Heubner as very potent "collateral systems" in emergency in addition to that of the circle of Willis and the extracranial anastomoses.

Since then the *extra*-cranial vessels have been investigated and their significance has been brought to our attention by Yates and Hutchinson (1961). Moreover, the demonstration of the four main cerebral arteries by neuroradiology has been adopted in routine investigation and is immensely valuable in clinical diagnosis and understanding of "strokes".

Meanwhile, the correlation of morphological studies, the analysis of the hemodynamics by the schools of Heymans and of Opitz and M. Schneider and the neuroradiological observations of clinical data have given an insight into the pathogenesis of the infarcts (Zülch, 1953—70), which may also facilitate the understanding of the clinical pathogenesis of strokes.

We are, however, far from a general consensus in this context. The dissention even seems to grow as the reader will notice in the various contributions. Three major concepts try to explain regional cerebrovascular insufficiency:

1. The hemodynamic theory, that diminished blood flow or infarction may result from acute occlusion or gross stenosis of a blood vessel when a fall in blood pressure occurs in the peripheral circulation. This concept came to the foreground by Denny-Brown (1951, 1960, 1965) when he tried to oppose the generally prevalent opinion that "angiospasm" would explain the diminished blood flow particularly in temporary attacks ("angiospastic insult" of Friedrich Kaufmann 1925, 1926 and v. Bergmann 1932).

2. A second concept has been propagated mainly by Kendell and Marshall (1963) that hypertension is the leading factor which introduces the pathological process in the majority of cases of stroke.

3. The interpretation of Ross Russell (1961) that cerebrovascular insufficiency may be the result of migration of microemboli from ulcerating thrombi has also to be considered.

All these concepts will be extensively discussed in this volume. This may suffice as a background for a closer discussion of the local factors impeding the lumina of vessels by stenosis or occlusion, usually on the basis of arteriosclerosis. This has been described by pathologists for a long time and even the significance of extra-cranial occlusion of the carotid artery (see Chiari, 1905; Dei Poli and Zucha, 1940; Hultqvist, 1942) was known before arteriography (Moniz, 1940; Riechert, 1943) established its clinical importance. Yates and Hutchinson's (1961) study gave a better understanding of the haemodynamic pattern of the resulting infarcts. In surgical ligation of the carotid artery, infarction of the brain has always been feared (Zülch and Herberg, 1949).

Having summarised the above views, the reader will estimate better the value of the following contributions of the various authors.

The chapter of Lhermitte and Gautier starts the discussion on a broad scale with a picture of *carotid occlusion* as a clinical entity. Pathology and pathogenesis serve as a basis to discuss the main theories of the pathogenesis of cerebrovascular insufficiency. They uphold the concept of "microembolisation" as a predominant cause and find but little evidence for the haemodynamic interpretation which has the "hypotensive crisis" as the heart of the pathological process. However, we shall find the view of "hypotensive crisis" defended by other contributors. The details of the syndrome of carotid occlusion make the first contribution so important for the clinician.

Their study is supplemented by Loeb's chapter on the clinical syndromes in the *vertebro-basilar* system, a concise contribution, which is a continuation of his more extensive work with J. S. Meyer (1965) on the same topic. Interestingly enough, Loeb stresses the discrepancy between the various "crossed" syndromes so well known from the classics and the actual pontine and mesencephalic vascular syndromes seen now. Loeb, from his observations, discusses first the definition of cerebrovascular insufficiency, which he regards as temporary, then defines its pathogenesis, where the local impedance may lie anywhere between the subclavian artery and the terminal branches of the basilar system.

To complement this clinical and morphological picture, Metzinger and Zülch discuss the typical patterns of the actual infarcts seen in a large series of vertebro-basilar thrombosis or stenosis. The types of infarct at the various levels—medulla oblongata, pons, cerebellum, mesencephalon, thalamus, hippocampus and occipital lobe are described. Here the lack of information regarding the exact and typical vascular supply of the various regions and the finer anastomoses and collateral pathways becomes obvious. This is in contrast with the supratentorial space where most of the arterial variations are well known from angiography. However, angiography provides much less information concerning the posterior fossa, especially regarding the midline and paramedian arteries of the brain stem, whose rostral patterns are much better known by visualisation of the carotid system.

The detailed studies of the classical period when the French, English, Italian and German anatomical schools provided fundamental information must be continued, particularly regarding the statistics of the variations of the vertebro-basilar system.

A study, which is so essential, is given by KRIBS and KLEIHUES on the recurrent artery of HEUBNER, which to date is rarely visualised radiologically and its clinical syndrome is almost unknown; it could not be detected even by this study. Since the time of the "great" and classical morpho-neurological syndromes have been discussed, finer vascular studies and the new interpretations of pathogenesis have led to an understanding of the new patterns of cerebral softenings or infarcts,which up to then were either unknown or unanalysable. The concept of the "last field" (M. SCHNEIDER, 1953) or the "frontier zones" (LINDENBERG and SPATZ, 1940) gave a better understanding both of the necroses of the spinal cord ("spinovascular insufficiency", ZÜLCH, 1954, 1967, 1970; critical zones in C4, D4 and L1) and also of the brain and were described as arising from such vascular insufficiency in the "terminal" ("last field") or "frontier zones" (1961).

Everybody present at the "Colloque de Marseille à Cologne" in 1964 will remember the final discussion on the morphological and electrobiological correlations of such infarcts or infarcted zones. The "parieto-temporo-occipital syndrome" in the "Dreiländereck" was born in this integrated discussion, where morphological ("Dreiländereck-Infarct", J. E. MEYER, 1958; ZÜLCH, 1961) angiographical (K. DECKER, 1958) and experimental (NAQUET, 1965, 1966) as well as clinical observations were correlated in GASTAUT's final discussion. This story is emphasized here, because we think that far more syndromes may arise in a similar manner. The two contributions of H. GASTAUT and his collaborators discussing frontier syndromes well documented by morphological, microangiographical and electrobiological findings speak for themselves as examples of an original method of analysis.

Such correlated morphological and macro- and micro-angiographical studies supplemented by regional blood flow measurements characterize the line of research of the team of G. LAZORTHES. It is interesting to see, how stone by stone the morphological structure of the vasculature is built and how this static research is automatically changed into more dynamic studies by the stimulation of the clinical and neurosurgical problems. And finally, the metabolic problems come into play rounding off the picture of cerebrovascular insufficiency both in vascular and in neoplastic disease.

The *second part* of this volume continues these contributions which focussed on morphology and integrated with related problems by studies of the anatomy of in vivo vessel-angiography. That this part has been made short, permits some introductory words on the indications and possible results of this technique.

Neuroradiology has become an important aid in the diagnosis of cerebro-vascular insufficiency by angiography of the four cerebral vessels. The concept of COHNHEIM (1872)—that the cerebral arteries were end-arteries—simplified for a long time the interpretations of cerebrovascular pathology so that in the clinical jargon arterial "thrombosis" and "brain softening" were used as synonyms. HEUBNER's observation of the reticular pattern of the cerebral arteries at the surface interconnected by "meningeal anastomoses" had not been generally accepted, though many morphologists and clinicians had observed contradictions to COHNHEIM's concept. They observed patent arteries with an infarct in the area supplied and occluded vessels without a related infarct because of the action of "HEUBNER's meningeal anastomoses" (LECENE and LHERMITTE, 1920). The rediscovery of these meningeal anastomoses (VAN DER EECKEN and ADAMS, 1953) was the first step towards a modern concept of the pathogenesis of infarcts. The next was the demonstration of a lack of arterial occlusions in the angiograms of patients with strokes (TATELMAN, 1958) in a high percentage of cases even after we learned to extend the examination proximally as far as the aortic arch. A haemodynamic interpretation of cerebrovascular insufficiency had to be offered. Moreover, the systematic studies of the neurologists and neuroradiologists working together with vascular surgeons led to the discovery of an astounding number of different and very complex individual or stereotyped anastomoses between the extra- and intracranial segments of the "four cerebral arteries".

FIELDS has given a bird's eye view of this very complex subject. It is regrettable that it was not possible to provide him with enough space for a demonstration of all the different collateral channels. This is partly mitigated by the fact that an extensive atlas of his observations has just been published (WEIBEL and FIELDS, 1969). He wanted, therefore, to discuss here only briefly the technique of cerebral angiography and visualisation of the most important collateral channels. Were there not the enormous variability in the morphological pattern of these extra- and intracranial arterial anastomoses, the resulting pattern could be predicted—as one very often can—in the cases of occlusion of a particular artery.

If one utilizes the pattern of the arterial supply revealed neuroradiologically to interpret the vascularisation of an area which is insufficiently supplied, one has to introduce the rules of haemodynamics. One must also interpret the pattern of the collateral supply dynamically and try to understand the actual flow in the arterial system. Here, an analysis of the various patterns which result, when anastomoses and other "detour" pathways are opened or utilized, may be of some help to understand the actual circumstances in this and the surrounding zone. My contribution attempts to demonstrate some patterns in the formation of anastomoses and collateral channels.

We have to further and describe the work of the neuroradiologist here more systematically. The neuroradiologist is able to analyse "cerebrovascular insufficiency" of the extra- and intracranial vessels from two points of view: a) from the *morphological* standpoint by showing congenital anomalies or occlusions, stenoses, ectasias and other alterations due to disease of the arterial wall and he may study it and b) from the *functional* standpoint by showing the *general circulation time* and also the *local changes of flow* in *time*. The morphological descriptions can be amplified in their value for the clinician when interpreted by the main theories of pathogenesis. Thus, neuroradiology participates in the process of diagnosis. We repeat briefly and concisely the three concepts:

1. The hypotensive crisis.
2. The hypertensive crisis.
3. The blood-platelet-fibrin embolisation.

as causes of cerebrovascular insufficiency.

The concept of a *hypotensive* crisis presupposes an impedance to regional blood flow of an artery, so that lowering intraarterial pressure causes a decrease of local flow. In terms of neuroradiology, the task is to determine and depict any "local" obstacle to flow such as a stenosing plaque, an occlusion by arteriosclerosis and/or thrombus which may be single or multiple. Also, the detecting of any "abnormal" collateral circulation is important, which if present, indicates an obstacle to flow which may escape local visualisation by neuroradiology. The same is true when *"regional retardation of blood flow"* becomes visible in serial angiography, if local alterations of the arterial lumina are not yet technically detectable. The sites of predilection of arteriosclerosis—so well known to the morphologist—need particular attention in the analysis of the angiogram. Finally, any vascular changes in a zone of infarction or cerebrovascular insufficiency or its surroundings may be demonstrated as hyperemia ("blush") and early ("red") filling veins (CRONQVIST, 1968). This will be dealt with below.

The *second theory* is that of arterial impedance by *"angiospasm"*, a concept which is now adapted to the clinical concepts under the term of "hypertensive crisis" (ZÜLCH, 1957, 1962).

Angiospasm, as is well known, can be seen in larger arteries after bleeding from saccular aneurysms and, for instance, in the carotid artery after puncture. Yet, as a cause of stroke, it will generally escape neuroradiological investigation because it is so transient, although CORDAY (1953, 1957) thought it could be detectable at the large arteries near the circle of Willis.

The *third theory*, that of arterial "microembolisation", is not of great importance radiologically, since alterations in the "local" circulation time may not be detectable, because such microembolisation will be so transient that it cannot be visualised.

The analysis of the pathological process may be evident by angiography and helps in understanding the pathogenesis. The actual circulation, however, is only partly evident by demonstrating the lumen of the vessels, the circulation time and by some special pathological features such as "hyperaemia" ("blush") and "early vein".

Therefore, one has tried to measure directly the blood flow for a long time and this will form the subject of the *third part of this book*. We are fortunate that Dr. S. KETY, who started this type of investigation experimentally and clinically, has provided a concise description of the historical development of the methods to solve these problems. He describes the results available at the present time. KETY and SCHMIDT (1948) developed their ingenious nitrous oxygen method after the principle of FICK and were the first to measure the blood flow in one hemisphere and to define a formula for the local circulation which is still the standard basis of comparison. This masterpiece of a short, exact, didactic paper leads the reader to the actual and most discussed methods of measurement, the regional cerebral blood flow (RCBF) measurements.

D. INGVAR again has personally been responsible for a good deal of progress in this subject. After his very clear chapter the reader will be familiar with the most modern results, the problems, pitfalls and the progress reached in the past few years. These methods described and discussed in many symposia are a very great advance in the experimental analysis of circulation and its disturbances and also show that the Xenon clearance measurement begins to have a place among bedside diagnostic procedures.

C. FAZIO and his collaborators discuss some very complex phenomena observed during these regional cerebral blood flow measurements, the haemodynamics of which are not yet entirely clear. Already angiography has shown some complicated "detours" of the circulation after occlusion of the subclavian arteries, which in clinical jargon have been called "subclavian steal". FAZIO and many others (LOEB and FAVALE, 1962) observed similar phenomena in the intracranial circulation, where in a "healthy" region a decreased circulation is observed because of a primary shift of blood to opposite cerebral areas, which need supplementary supply. The authors develop a theory explaining these exciting hemodynamic phenomena of "steal" or "counter steal"; the above described phenomenon they call the "altruistic attitude" which is detrimental to the donor hemisphere.

When we try to introduce these most important observations and phenomena into the biological analysis of an area with cerebrovascular insufficiency or an infarct, we still do not understand fully their significance.

If we return to the theoretical concept of the pathogenesis of an infarct, nobody could predict what "loss of autoregulation", "luxury perfusion", "steal" and "countersteal" mean in terms of a qualitative and quantitative damage to the tissue or how the most important question, that of restitution "ad integrum", morphologically and functionally can be answered.

We have already some data as to the duration of the ischemia and the time of recovery based on animal experiments and observations in man. Up to now there were only minor discrepancies in the first "survival times" published by WEINBERGER, GIBBON and GIBBON (1940, 3 to 4 min) and M. SCHNEIDER (1964) and HIRSCH et al. (1957, up to 8 min). All this seems important at a time when the "definition of death" is so basic because of the question of transplantation of organs. Yet, it seems as if these accepted figures are likely to be rediscussed.

K. A. HOSSMANN and Y. OLSSON have succeeded in extending this time of complete ischemia to 30 min (!) and by experiments of HOSSMANN and SATO (1970) to 1 hr. By an interesting method, namely, evoked potentials, they show that the nervous tissue is apparently not more vulnerable than the mesoderm of the capillaries and it is probably the "no-reflow" phenomenon which prevents restitution of function after the return of the circulation. When completed, these studies may become of uttermost importance, particularly when the biochemical processes involved in the closing of the capillaries are understood and can be prevented or counteracted. This may become an important progress in bedside therapy.

At this point of our discussion we must interpret the value of all the methods referred to in these contributions for a biological evaluation of the nervous tissue in cases of regional cerebrovascular insufficiency. Up to the present neither the neurological examination nor any of the diagnostic techniques such as EEG or scanning, nor any of the data of clinical pathology can predict the viability, that is the state of reversibility of the tissue damage. The observations of OPITZ and M. SCHNEIDER and his school are still valid, namely that a decrease in flow by 50% leads to functional damage, but only a lowering to 20% of the normal circulating blood or 15% of the available oxygen provokes necrosis. A circulation in this "safety zone" of one third of the values stated above suffices to prevent infarct; yet by clinical means, even by the regional cerebral blood flow measurements, we cannot quantitate the circulation sufficiently. Circulation is not a static unchanging factor, but changes as much as the blood pressure.

This problem of safeguarding the necessary blood pressure level seems to us one of the important problems. We have to return to this when we introduce the contributions on hypo- and hypertension as factors in cerebrovascular pathogenesis.

How rapid changes may be, is shown in the article of J. S. MEYER and his collaborators. His prolonged studies have developed models to show such rapid changes not only in the circulation, but also in the composition of the blood, which we did not so far introduce into our analysis of the regional disturbance of flow. They can follow rapid changes in the blood gases and the pH etc. in man. These data can be correlated to other clinical symptoms. Moreover, interesting changes in the conditions of arousal from sleep and during "rapid eye movement" (REM) sleep are discussed. Here an entirely new field of investigation of clinical pathology is opened, which may one day help in predicting the viability of the brain tissue discussed above as in analogy to the transaminases in coronary infarct. One day we may be able to predict by finer data obtained by clinical pathological methods whether there is only a transient hypoxic change or if there is already oedema or necrosis in the tissue.

Other changes in the blood flow which are more "morphological" are discussed by Swank and Bartsch on a broad biological background. Their screen filtration method allows a constant control of the viscosity; perhaps some of our problems in cerebrovascular insufficiency are based on such phenomena as microaggregation in the endangered zone or perhaps even throughout the general circulation. The treatment with low molecular dextrans, so extensively and sometimes successfully used, may be due to this phenomenon. More and more facets in the very complex picture of regional cerebrovascular insufficiency thus accrue.

The *final part* of this presentation of cerebrovascular problems deals with more clinical questions, but also with problems of pathogenesis. The discussion of Marshall is of great interest as to how far hypertension is the cause of cerebrovascular changes. The pathogenesis of mass hemorrhages on the basis of the famous "aneurysms" (of Charcot and Bouchard, 1868, of Pick, 1910, and Ellis, 1909, in classical times) and furthermore the clinical phenomena of hypertensive arterial disease are considered as well as their importance in the evolution of infarcts. The origin of these aneurysms and the correlation between hypertension and arteriosclerosis are rediscussed and the question raised as to how often the lacunae are micro-infarcts due to occlusion of hyalinised arteries and if the bleeding around such arteries may actually be the cause of transient episodes. Since hypertension is controlled better than some 30 years ago, the medical care of hypertensive patients is a serious obligation to medical neurology.

Katsuki and his collaborators' contribution has a great statistical background and may be interesting not only as such, but also may have some bearing on "geographical" (environmental) or racial pathology. They assessed data concerning age, blood pressure and diabetes and correlated these with infarct or stenotic lesions detectable by angiography. These are positively correlated to age, but not, or less strictly, correlated to the two different forms of hypertension and to diabetes mellitus. The data presented, provided that they are not peculiar to the Japanese population will be of great value for any similar statistics. The great discrepancies in data provided by environmental factors we have just learned to appreciate by the world-wide study of the Minnesota group which stimulated many others (A. B. Baker et al., 1960, 1967; Resch et al., 1965, 1967; Williams et al., 1969 etc.).

Finally, Ross Russell's paper introduces his concept of microembolisation as the cause of the transient ischemic attacks. He makes the reader familiar with the pathophysiology of the platelets and ends by a summary of his interesting theses which support his views. Logically he deduces the medical regime which could prevent or counterattack this process of embolisation.

As a contrast, PIERACH's contribution is focussed on "relative" hypotension as one of the main factors in the haemodynamic disturbance leading to temporary or permanent neurological damage. His point is that it is the "relative" drop of blood pressure, i.e. for instance in a hypertensive patient to "normal" or slightly subnormal values, which induces the regional decrease of flow similar to the interpretations in some types of hypoglycemia. Clinical evidence is given for such a concept.

That blood pressure indeed is only a "relative" and unsteady value is convincingly shown by VOLKER HOSSMANN (1969). It is a factor which has some "genuine" tendencies, but is dependent on environmental and also particularly "circadian" factors. It can only be expressed as a "tendency toward" a certain figure. For a long time, in our clinical studies the fluctuations of the blood pressure values were visible with certain drops and increases in the 24 hr-spectrum. Blood pressure seems to undergo the same rhythmic changes as other similar factors in the autonomic nervous system. This has been proven in a series of normal patients ("lows" at 3 a.m. and less marked in the early afternoon). HOSSMANN makes clear that these drops of blood pressure, alone or as an additional factor, may serve to induce cerebrovascular insufficiency and based his interpretation on a large series of patients with strokes.

One of the examples of such a "midnight" crisis of blood flow we detected in the first description of carotid endarterectomy by EASTCOTT, PICKERING and ROB (1954) and it is worthwhile reading it in the original.

A patient of 52 had, altogether, 33 attacks of tachycardia occurring always at 2 o'clock in the morning with angina pectoris of a few minutes duration and with difficulty in breathing and an acute drop in the blood pressure. A few minutes after each of these attacks paresis of the arm and signs of aphasia occurred, as well as amaurosis of the left eye. These symptoms disappeared after 30 min. His blood pressure was often 240/150 and the left heart was correspondingly enlarged.

Angiography showed carotid stenosis. This narrowing of the vessel encouraged the authors to postulate that the disturbance of blood flow to the brain had been haemodynamically produced by stenosis. This was the indication for the resection which gave the green light for the future of operative endarterectomy. After the operation the patient had no further *transient ischemic attacks*. What was interesting, was that the attacks of tachycardia and angina pectoris and their predilection for occurring at 2 o'clock in the morning remained.

PICKERING had sought to lower the blood pressure in this patient to 124/68 but he dared not risk this for longer than 15 sec. He did not achieve in this way any production of the neurological symptoms, but he concluded that the drop in blood pressure was too short. He could not risk prolonging it further because of the risk of endangering the patient.

However, the authors emphasize the good results of surgical removal of the stenosis with the subsequent absence of transient neurological attacks as providing support for their haemodynamic explanation. They thought that one would have to strain one's imagination to assume an "angiospastic" basis to explain 33 different attacks without any specific cause which all affected the arteries of the left eye and the left cerebral hemisphere and caused sufficient constriction to produce focal ischemia, while all the other arteries remained free from this angiospastic occurrence. However, the nocturnal angina pectoris attacks persisted! It would be difficult to interpret this as attacks of microembolisation.

The reader will have noticed that the "hypotensive crisis" seems to me to be still one of the most potent factors in introducing regional cerebrovascular insufficiency. Its pathogenesis seems to be a complex mechanism and I do not believe that hypotension is the sole cause. I believe that in man in such an attack—the hallmark of which is a fall in blood pressure—far more changes are set into action than simply the "diminished blood pressure" and perhaps the above mentioned disturbance of viscosity acting on microcirculation.

The logical conclusion is that one must postulate a "multiple factor theory" as I have exemplified on many occasions by describing numerous case histories (1962). Pathogenesis of regional or general cerebrovascular insufficiency must be studied by all of us with the greatest energy since its knowledge is the only solid basis of treatment.

To come back to our discussion of the diagnostic possibilities for a proper evaluation of pathogenesis, it may by emphasized how little electroencephalography has contributed to date to our knowledge. It may, however, rapidly show the changes of deficient circulation. It shows also gross ischemic changes in seconds but reacts only slowly to minor changes of circulation, if they do not lead to edema, i.e. a disturbance of the blood-brain-barrier.

One had hoped that isotope scanning would be more sensitive, but its results up to now are very controversial. Our own results (Marx and Rosarius, 1969) with 99m Technetium (the "agent of choice", Webber, 1965) are similar to the results of other authors.

The method is not troublesome to the patient. It should be performed around the 3rd to 5th day along with angiography and may be repeated around the 21st day. The pathophysiological basis is still controversial, but may have something to do with disturbances of the blood-brain-barrier.

The percentage of positive scans in cerebrovascular insufficiency was reported as 90% by Ojeman et al. (1964, 1966), Croll et al. (1968) and Glasgow et al. (1965, 1967) and as 17% by Brown et al. (1967) in ischemic, non-hemorrhagic lesions. This latter corresponds to our own figure (1969, 9 positive scans in 54 cases). The reliability of this procedure is very low for cerebrovascular insufficiency cases.

We had, for instance, two similar cases of occlusion of the middle cerebral artery but only one was positive; we were not able to determine the criteria for "positivity" or to make any conclusion regarding the prognosis on this method.

Since we cannot predict the viability, we have to continue energetic treatment for weeks until clinical failure indicates that the tissue is infarcted and functional recovery cannot be expected. This is a very unsatisfactory situation for the patient, the doctor and also from an economic standpoint.

What is the program of our medical treatment of "strokes"? I can answer this question only according to my personal interpretation of pathogenesis in the various types of "stroke" and from the experience of other clinicians.

The lines of our treatment are:

1. Restoration of the local cerebral blood flow by restoration of the systemic blood pressure. This must be considered for each particular patient, i.e., taking into account the normal tendencies of his blood pressure.

2. Safeguarding of the adequacy of the general circulation, i. e., stabilization of the heart's action.

3. Combating the effects of hypoxia, particularly cerebral edema and disturbances of the blood-brain barrier.

4. Elimination of disturbances of respiration (oxygenation, CO_2 exhalation), blood viscosity (microcirculation), disturbances of inorganic metabolism, hydration and renal function.

This may induce the following more detailed treatment programme (in collaboration with W. GEHLEN, head of our intensive care unit). The patient ought to get to the intensive care unit as soon as possible. The steps to be followed in diagnosis are:

A. Examination of consciousness, circulation and particularly of heart and blood pressure, of neurological deficit and bodily state. Gross tests for diabetes and renal function, ECG, ultrasound diagnosis, spinal tap etc. Permanent control of pulse, breathing and blood pressure.

B. Treatment programme for heart and peripheral circulation (blood pressure): glycosides regulating heart action, stabilisation of blood pressure to "normality", i.e. the different values according to state before stroke in the hyper-, normo- or hypotensive patient. So-called "vasoactive" drugs to restore brain circulation: aminophylline, dehydrated ergotamines (hydergine or dihydroergotamine according to blood pressure). Regulation of water balance, mineral and energy-household; antioedematous treatment: infusion therapy with low molecular dextran and glucose, or sorbitol, mannitol, lasix, diamox; occasionally cortisone; anti-infectious, anti-decubitus treatment. If necessary: intubation, artificial respiration or assisted breathing, tracheostomy.

This programme is the first step in the treatment of the patient with a "stroke". When the patient's state permits, the second phase of a more exact and finer medical examination and EEG follow and angiography is performed, which ought to be as complete as possible at this stage of the disease; furthermore, scanning and, if possible, RCBF measurement. This must suffice as a framework of a medical programme.

Interesting results have been obtained particularly in the prevention of strokes by vascular surgery. This important question is very precisely discussed by E. J. WYLIE's paper. He reports indications for surgery in the various types of cerebrovascular insufficiency, in the transient ischemic attack and in "acute stroke" as well as in "chronic" stroke and the "chronic low perfusion syndrome". He also deals with the indications in asymptomatic cases. Since his attitude is in general conservative, it will convince the even more conservative non-surgical clinicians of this operative programme. Perhaps the final indications for surgery will emerge only after the great American randomised statistics become available. However, they may not deviate profoundly from Dr. WYLIE's thoughts.

References

BAKER, A. B., FLORA, G. C., RESCH, J. A., LOEWENSON, R.: The geographic pathology of atherosclerosis: A review of the literature with some personal observations on cerebral atherosclerosis. J. chron. Dis. 20, 685—706 (1967).
— REFSUM, S., DAHL, E.: Cerebrovascular disease. IV. A study of a Norwegian population. Neurology (Minneap.) 10, 525—529 (1960).
BERGMANN, G. v.: Funktionelle Pathologie. Kap. 12: Die Lehre von der Apoplexie. Berlin: Springer 1932.

Brown, A., Zingesser, L., Scheinberg, L. C.: Radioactive mercury-labeled chlormerodrin scans in cerebrovascular accidents. Neurology (Minneap.) 17, 405—412 (1967).

Büchner, F.: Die Coronarinsufflzienz. Dresden: Steinkopff 1939.

— Spezielle Pathologie, 4. Aufl. Wien: Urban & Schwarzenberg 1965.

— Allgemeine Pathologie, 5. Aufl. Wien: Urban & Schwarzenberg 1966.

Charcot, J. M., Bouchard, Ch.: Nouvelles recherches sur la pathogénie de l'hémorrhagie cérébrale. Arch. Physiol. Paris 1, 110—127, 643—675, 725—734 (1868).

Chiari, H.: Über das Verhalten des Teilungswinkels der Carotis communis bei der Endarteriitis chronica deformans. Verh. dtsch. Ges. Path. 9, 326 (1905).

Corday, E., Rothenberg, S. F.: The clinical aspects of cerebral vascular insufficiency. Ann. intern. Med. 47, 626 (1957).

— — Putnam, T.: Cerebral vascular insufficiency. An explanation of some types of localized cerebral encephalopathy. Arch. Neurol. Psychiat. (Chic.) 69, 551—570 (1953).

Croll, M. N., Brady, L. W., Faust, D. S., Kazem, I., Antoniades, J., Tatem, H. R.: Comparison brain scanning with mercury 203 and technetium 99m. Radiology 90, 747 to 749 (1968).

Cronqvist, S.: Regional cerebral blood flow and angiography in apoplexy. Acta radiol. Diagn. 7, 521—534 (1968).

Decker, K.: Der Schlaganfall als neuroradiologisches Problem. Dtsch. med. Wschr. 83, 205 (1958).

Dei Poli, G., Zucha, J.: Beiträge zur Kenntnis der Anomalien und der Erkrankungen der Arteria carotis interna. Zbl. Neurochir. 5, 209—238 (1940).

Denny-Brown, D.: The treatment of recurrent cerebrovascular symptoms and the question of „vasospasm". Med. Clin. N.Amer. 35, 1457—1474 (1951).

— Recurrent cerebrovascular episodes. Arch. Neurol. Psychiat. (Chic.) 2, 194—210 (1960).

— Discussion. Proc. Aust. Ass. 3, 100 (1965).

Eastcott, H. H. G., Pickering, G. W., Rob, C. G.: Reconstruction of internal carotid artery in a patient with intermittent attacks of hemiplegia. Lancet 1954 II, 994—996.

Ellis, A. G.: The pathogenesis of spontaneous cerebral hemorrhage. Proc. path. Soc. Philad., Vol. 12 (N.S.) 3, 197 (1909).

Gastaut, H., Naquet, R.: Etude électroencéphalographique de l'insuffisance circulatoire cérébrale. Symposium international sur la circulation cérébrale, p. 163—191. Sandoz 1965.

Glasgow, J. L., Currier, R. D., Goodrich, J. K., Tutor, F. T.: Brain scans at varied intervals following C.V.A.-J. nucl. Med. 6, 902—916 (1965).

— Brain scans of cerebral infarcts with radioactive mercury. Radiology 88, 1086 (1967).

Heubner, O.: Die luetische Erkrankung der Hirnarterien. Leipzig 1874.

Heymanns, C.: Regulation of blood pressure and hypertension. Trans. Hunter. Soc. 25, 38—45 (1967).

Hirsch, H., Euler, K. H., Schneider, M.: Über die Erholung und Wiederbelebung des Gehirns nach Ischämie bei Normothermie. Pflügers Arch. ges. Physiol. 265, 281—313 (1957).

— Schneider, M.: Durchblutung und Sauerstoffaufnahme des Gehirns. Handb. Neurochir., Bd. I/2, S. 434—552. Berlin-Heidelberg-New York: Springer 1968.

Hossmann, K.-A., Sato, K.: Recovery of neuronal function after prolonged cerebral ischemia. Science 168, 375—376 (1970).

Hossmann, V.: Der Hirninfarkt und seine Abhängigkeit von exogenen und endogenen Faktoren des Blutdrucks. Doktor-Diss., Universität Köln 1969.

Hultquist, G. T.: Über Thrombose und Embolie der A. carotis. Jena: Fischer 1942.

Kaufmann, Fr.: Klinisch-experimentelle Untersuchungen zum Krankheitsbild der arteriellen Hypertension. Z. exp. Med. 42, 473—495; 43, 141—169 (1925); — Z. klin. Med. 100, (1925).

— Über Blutdruckschwankungen und ihre Bedeutung für den Organismus. Hypertension. Ärztl. Fortbildungskurs Bad Nauheim. Leipzig: Thieme 1926.

Kendell, R. E., Marshall, J.: Role of hypotension in the genesis of transient focal cerebral ischaemic attacks. Brit. med. J. 5353, 344—348 (1963).

Kety, S. S., Schmidt, C. F.: The nitrous oxyde method for the quantitative determination of cerebral blood flow in man: theory, procedure and normal values. J. clin. Invest. 27, 476—483 (1948).

LECENE, B., LHERMITTE, J.: Une observation anatomo-clinique d'un cas de ramollissement cérébral consécutif à l'obliteration de l'artère sylvienne gauche par embolie métallique. Rev. neurol. **27**, 1116—1121 (1920).

LINDENBERG, R., SPATZ, H.: Über die Thromboendarteriitis obliterans der Hirngefäße (cerebrale Form der v. Winiwarter-Bürgerschen Krankheit). Virchows Arch. path. Anat. **305**, 531—557 (1940).

LOEB, C., MEYER, J. S.: Strokes due to vertebro-basilar disease. Springfield (Ill.): Charles C. Thomas 1965.

LOEB, E., FAVALE, F.: Contralateral EEG abnormalities in intracranial arteriovenous aneurysms. Arch. Neurol. Psychiat. (Chic.) **7**, 121—128 (1962).

MARX, P., ROSARIUS, C.: Angiographische und hirnszintigraphische Befunde beim Hirninfarkt. Radiologe **9**, 428—431 (1969).

MEYER, J. E.: Zur Lokalisation arteriosklerotischer Erweichungsherde in arteriellen Grenzgebieten des Gehirns. Arch. Psychiat. Nervenkr. **196**, 421—432 (1958).

MONIZ, E.: Die cerebrale Arteriographie und Phlebographie. Ergänzungsband II zum Handbuch der Neurologie von Bumke-Foerster. Berlin: Springer 1940.

NAQUET, R., ARFEL, G., CHOUX, M., DUBOIS, D.: Etude expérimentale de l'embolie gazeuse par voie carotidienne chez le chat. Electroenceph. clin. Neurophysiol. **20**, 181—196 (1966).

— FRANCK, G., VIGOUROUX, R.: Données nouvelles sur certaines décharges paroxystiques du carrefour parieto-temporo-occipital rencontrées chez l'homme. Zbl. Neurochir. **25**, 153—180 (1965).

OJEMANN, R. G., ARONOW, S. A., SWEET, W. H.: Scanning with positron-emitting radioisotopes. Arch. Neurol. (Chic.) **10**, 218 – 232 (1964).

— — — Scanning with positron-emitting isotopes in cerebrovascular disease. Acta radiol. Diagn. **5**, 894—905 (1966).

OPITZ, E., SCHNEIDER, M.: Über die Sauerstoffversorgung des Gehirns und den Mechanismus von Mangelwirkungen. Ergebn. Physiol. **46**, 126—260 (1950).

PICK, L.: Über die sogenannten miliaren Aneurysmen der Hirngefäße. Berl. med. Wschr. **47**, 325—329 (1910).

REIN, H.: Die Physiologie der Coronardurchblutung .Verh. dtsch. Ges. inn. Med. Wiesbaden München: J. F. Bergmann 1931.

RESCH, J. A., BAKER, A. B.: An epidemiological approach to the study of cerebrovascular disease. Med. Tms (N.Y.) **1**, 9 (1965).

— OKABE, N., LOEWENSON, R., KIMOTO, K., KATSUKI, S., BAKER, A. B.: A comparative study of cerebral atherosclerosis in a Japanese and Minnesota population. J. Atheroscler. Res. **7**, 687—693 (1967).

RIECHERT, TR.: Die Arteriographie der Hirngefäße, 1. Aufl. München-Berlin: J. F. Lehmanns-Verlag 1943.

ROSS RUSSELL, R. W.: Observations on the retinal blood vessels in monocular blindness. Lancet **1961 II**, 7218, 1422—1428.

ROTHENBERG, S. F., CORDAY, E.: The etiology of the transient stroke. J. Amer. med. Ass. **164**, 2005 (1957).

SCHNEIDER, M.: Die Wiederbelebungszeit verschiedener Organe nach Ischämie. Arch. klin. Chir. **308**, 253—265 (1964).

— Durchblutung und Sauerstoffversorgung des Gehirns. Verh. dtsch. Ges. Kreisl.-Forsch. **19**, 3—25 (1953).

TATELMANN, M.: The angiographic evaluation of cerebral atherosclerosis. Radiology **70**, 801—810 (1958).

VAN DER EECKEN, H., ADAMS, R. D.: The anatomy and functional significance of the meningeal arterial anastomoses of the human brain. J. Neuropath. exp. Neurol. **12**, 132—157 (1953).

WEBBER, M. M.: Technetium 99m normal brain scans and their anatomic features. Amer. J. Roentgenol. **94**, 815—818 (1965).

WEIBEL, J., FIELDS, W. S.: Atlas of arteriography in occlusive cerebrovascular disease. Stuttgart: Thieme 1969.

WEINBERGER, L. M., GIBBON, M. H., GIBBON J. H., JR.: Temporary arrest of the circulation to the central nervous system. Arch. Neurol. Psychiat. (Chic.) **43**, 615 (1940).

WILLIAMS, A. O., RESCH, J. A., LOEWENSON, R. B.: Cerebral atherosclerosis — a comparative autopsy study between Nigerian negroes and American negroes and Caucasians. Neurology (Minneap.) **19**, 205—210 (1969).

YATES, P. O., HUTCHINSON, E. C.: Cerebral infarction: the role of stenosis of the extracranial cerebral arteries. London: Her Maj. Stat. Off. 1961.

ZÜLCH, K. J.: Mangeldurchblutung an der Grenzzone zweier Gefäßgebiete als Ursache bisher ungeklärter Rückenmarksschädigungen. Dtsch. Z. Nervenheilk. **172**, 81—101 (1954).

— Der cerebrale Insult. In: Die essentielle Hypertonie als Schädigungsfolge. Herbsttagung 1957 des Ärztl. Sachverständigen-Beirats für Fragen der Kriegsopferversorgung, Bonn.

— Gedanken zur Entstehung und Behandlung der Schlaganfälle. Dtsch. med. Wschr. **35**, 1524—1530; **36**, 1585—1590 (1960).

— Über die Entstehung und Lokalisation der Hirninfarkte. Zbl. Neurochir. **21**, 158—178 (1961).

— Die Pathogenese von Massenblutung und Erweichung unter besonderer Berücksichtigung klinischer Gesichtspunkte. Acta neurochir. (Wien) Suppl. VII, 51—117 (1961).

— Gedanken über die Entstehung der Hirninfarkte und anderer Durchblutungsstörungen. Kongreßband der I. Internat. Salzburger Konferenz (BERTHA, A., EICHHORN, O., LECHNER, H., Hrsg.), Graz 1962.

— Neuere Anschauungen über die Entstehung der cerebralen Insulte. Extrait du Livre du Dr. Ludo van Bogaert. Acta Med. Belgica, Brüssel, 890—904 (1962).

— Zur Pathogenese des cerebrovaskulären Insultes. Internist **4**, 64—70 (1963).

— Morphology and pathogenesis of cerebral infarction. Annales del XII Congreso Latinoamericano de Neurocirugia. Symposium Internacional de Investigaciones Neurologicas, p. 205—265. Lima 1967.

— Die spinale Mangeldurchblutung und ihre Folgen. Verh. dtsch. Ges. inn. Med. **72**, 1007—1059 (1967).

— Reconsiderations of the clinical problem of cerebrovascular insufficiency. In: Research on the cerebral circulation. III. International Salzburg Conference, p. 1—41 (MEYER, J. S., LECHNER, H., EICHHORN, O., Eds.). Springfield (Ill.): Charles C. Thomas Publ. 1969.

— Angiographische Befunde zur Pathogenese der Hirndurchblutungsstörungen. Zbl. Neurochir. **31**, 1—25 (1970).

— HERBERG, H.-J.: Das klinische Bild der akuten Blutsperre der Arteria carotis. Dtsch. Z. Nervenheilk. **160**, 38—79 (1949).

— KLEIHUES, P.: Neuropathology of cerebral infarction. Thule Intern. Symposia Stockholm 1966, p. 57—75. Stockholm: Nordiska Bokhandelns Förlag 1967.

— — GABE, D.: Die aktuelle Problematik auf dem Gebiet der Pathogenese, Klinik und Therapie der Hirndurchblutungsstörungen. II. Internationale Salzburger Konferenz 1964, S. 339—367. Wien: Verlag Brüder Hollinek 1964; — Wien. med. Wschr. **116**, 494—503 (1966).

— KURTH-SCHUMACHER, R.: The pathogenesis of "intermittent spinovascular insufficiency" ("spinal claudication of Dejerine") and other vascular syndromes of the spinal cord. Vasc. Surg. **4**, 116—136 (1970)

Clinico-Pathological Studies on Cerebrovascular Disease

Internal Carotid Artery Stenosis and Occlusion*
A Short Pathological and Clinical Survey

F. Lhermitte and J. C. Gautier

Internal carotid artery (I.C.A.) lesions are a major pathological condition in cerebral infarction or ischaemia. In a series of 432 largely unselected post-mortem examinations, Fisher (1954) found 218 cases with vascular lesions. Carotid disease was present 41 times (28 occlusions of one or both I.C.A.; 13 very severe stenoses of the arterial lumen), that is in 9.5%. The prevalence of advanced carotid disease in this series was about the same as that of cerebral haemorrhage and that of lacunes. In contrast, thrombotic occlusion of major vessels adjacent to the circle of Willis was found only 12 times. Lhermitte, Gautier, Derouesné and Guiraud (1968) found an I.C.A. occlusion in 45.9% and a significant stenosis in 9% of 122 cases of infarction or ischemia in the territory of the middle cerebral artery.

Aetiology of I.C.A. Stenosis and Occlusion
Embolic occlusion of the carotid artery is by no means rare. In the present paper, however, attention has been paid only to thrombotic occlusion. Three main aetiologies are briefly reviewed here (for a detailed study see Lhermitte, Gautier and Derouesné, 1966):

1. Trauma. Trauma is very frequent in the young in whom it is responsible for one quarter of I.C.A. thromboses (Pitner, 1966). Arterial lesions may result from: a) direct trauma upon the neck; b) trauma into the mouth behind the tonsil; c) bruising of the artery against the transverse process of the atlas when the neck is suddenly and forcefully extended. A dissecting aneurysm with subsequent mural thrombus resulting in embolism and/or occlusion appears as a likely mechanism of the cerebral accident in a number of such cases.

2. Arteritis. Arteritis of the I.C.A. mainly occurs in the child, in the course of ear and throat infections, cervical adenitis or after tonsillectomy (Shillito, Jr., 1964; Bickerstaff, 1964).

3. Atherosclerosis. This is by far the most important cause of I.C.A. stenosis and occlusion in the adult and aged.

a) Prevalence. Evidence is available from post-mortem studies. Yates and Hutchinson (1961) found 70 stenoses reducing the lumen by more than half among 100

---

* From the Hôpital de la Salpêtrière, Paris.

cases which had been selected upon a clinical diagnosis of cerebral ischaemic accident, excluding cases of cerebral infarction obviously not due to degenerative cerebro-vascular disease. MITCHELL and SCHWARTZ (1965) found the lumen reduced by more than half in 1 male out of 7, the entry into this series being only ages above 35. FISHER, GORE, OKABE and WHITE (1965) in 178 largely unselected cases recorded 17 stenoses reducing the lumen by more than 75%. In a series of 75 cases from a Department of Neurology, LHERMITTE, GAUTIER and DEROUESNÉ (1966) found 15 stenoses reducing the lumen by more than 75%.

I.C.A. occlusions were present in 7% of the cases of FISHER et al. (1965), 16% of those of YATES and HUTCHINSON (1961), 18.6% of those of LHERMITTE et al. (1966).

b) Sex. Age. In a post-mortem series, tight stenoses, i.e. > 75%, were 7 or 8 times more frequent in men than in women (LHERMITTE et al., 1966). Similarly, in a clinical and angiographic study of 107 patients (85 occlusions, 36 stenoses) 79.4% were men (ALAJOUANINE, LHERMITTE and GAUTIER, 1965). A similar predilection for coronary atherosclerosis is found in men (LENEGRE et al., 1961).

Tight stenoses and occlusions are mostly encountered during the 5th, 6th and 7th decades of life. In the series of ALAJOUANINE et al. (1965), mean age, depending to some extent on the prerequisite of angiography for entry, was 53 years 9 months. Mean age was significantly lower in men than in women, a difference not accounted for by the greater longevity of women. The same pattern according to age and sex is found in coronary atherosclerosis (LENEGRE et al., 1961). When all causes of I.C.A. lesions are considered together (LUESSENHOP, 1959) it appears that about one quarter of the cases have occurred before the age of 35, a figure which approximately indicates the prevalence of non atherosclerotic lesions.

c) Sites of lesions. Almost all tight stenoses develop on the proximal 20 mm of the artery (the sinus). LHERMITTE et al. (1966) found 13 stenoses > 75% on the sinus and only 2 on the terminal part of the artery (the syphon). The cervical part of the artery is usually free from significant lesions.

Occlusions resulting almost always from thrombus superimposed on or near a stenosis, most thrombi develop initially in the proximal part of the I.C.A. HULTQUIST (1942) found that occlusion had begun in the sinus in $2/3$ of his cases, in the syphon in $1/3$. In the series of FISHER et al. (1965) all occluding thrombi had originated in the proximal I.C.A. In 18 post-mortem cases of our own, we found that the thrombus had originated in the sinus in 11, in the syphon in 2, while in 5 it was deemed impossible to know for certain where thrombosis had actually begun.

Such data obviously emphasize the vital importance of atherosclerotic lesions of the proximal part of the I.C.A.

It is well known that atherosclerosis is a multifocal process which, moreover, shows a definite tendency to develop in particular sites (bifurcations, bends) in arterial systems. Therefore it is to be expected that significant atherosclerotic lesions are likely to involve both proximal I.C.A. LHERMITTE et al. (1966) found that when one sinus is stenosed by more than 75% the odds are $3/10$ that the other is as severely stenosed and only $1/10$ that it is devoid of significant lesions. MITCHELL and SCHWARZ (1965) have reported similar findings.

Such figures should be taken into account when contralateral angiography or surgical procedures are contemplated.

Some Mechanisms of Infarction and Ischaemia in Atherosclerosis

1. Haemodynamic Effects of I.C.A. Stenoses

These have been investigated in recent years by several techniques. The velocity of blood flow may be measured by arm-to-retina fluorescein time. Pressure in the I.C.A. may be inferred, at least compared with that in the contralateral artery, by ophthalmo-dynamometry. Blood flow has been measured by two independent groups of workers, (TINDALL, ODOM, CUPP JR. and DILLON, 1962; BRICE, DOWSETT and LOWE, 1964). In short it appears that pressure and blood flow are being reduced only when the arterial lumen is narrowed by 70 to 90%. Similarly, it would appear that a decrease of flow velocity occurs only when a tight stenosis is present (GILLAND, HELGASON, LEISSNER and STATTIN, 1965). Clinical experience supports these data, for beyond tight stenoses of the sinus, a very good filling of the arterial tree is usually seen on angiograms. Therefore, it appears that a moderate or even a tight stenosis should not be accepted as a simple explanation of a stroke and that some additional co-factor must be sought.

In this state of affairs, falls in blood pressure are traditionally called for. However, there is not really much convincing evidence to support the view according to which they do play a part in many patients (for a review, see LHERMITTE et al., 1966).

2. Formation and Evolution of Thrombi

For reasons rather unclear at present, thrombosis may occur in or near arterial stenoses. Present knowledge of this process has been reviewed by POOLE and FRENCH (1961), MITCHELL and SCHWARTZ (1965) and GAUTIER (1966). It appears that, first, platelets stick to the arterial wall, then they clump in a white mass, the white thrombus. Fibrin is absent or very scarce. Upon the white thrombus may superimpose a red thrombus, made of red cells entrapped in a mesh of fibrin. The thrombus may either result in a stenosis (mural thrombus) or in a complete block of the artery (occluding thrombus).

There is evidence that thrombi may break into pieces, resulting in emboli. Platelet emboli (from white thrombus) being small and devoid of fibrin are likely to occlude small arteries and to crumble more or less quickly, thus determining transient arterial occlusions. Emboli from red or mixed thrombus are likely to be of greater volume and, due conceivably to the presence of fibrin, more solid. Therefore they may occlude major arteries temporarily or definitively.

Occlusive thrombosis of the I.C.A. sinus is always, or almost always, followed by the development of an anterograde thrombus, the so-called stagnation thrombus, which extends up to the intra-petrous or intra-cavernous part of the artery (CASTAIGNE, LHERMITTE and GAUTIER, 1965; GAUTIER, 1966).

In some patients, the thrombus stops short of the ophthalmic artery, thus allowing collateral supply through the external carotid-ophthalmic by-pass and principally through the circle of WILLIS. Thus infarction may be escaped in a fair number of patients. HULTQUIST (1942) found cerebral lesions in only $1/3$ of the patients in whom the arterial occlusion stopped short of the ophthalmic artery and FISHER et al. (1965) found 5 asymptomatic cases among 15 cases of I.C.A. occlusion.

In many patients, however, the anterograde thrombus extends beyond the ophthalmic artery and beyond the circle of WILLIS into the cerebral arteries. Thus collateral supply is severely impaired and, moreover, the mouth of the arteries which penetrate the brain is occluded. Such an occlusion of the perforating arteries obviously is also at play when an embolus lodges in a cerebral artery beyond the circle of Willis. In our experience, severe permanent neurological deficit almost always results from a thrombotic or embolic occlusion beyond the circle of Willis.

Besides breaking into emboli or occluding the artery, thrombi may be covered by endothelial cells and embodied in the arterial wall. According to Duguid (1946, 1948, 1949), this process might end in atherosclerotic lesions.

Symptoms and Signs of Atherosclerosis, I.C.A. Stenosis and Thrombosis

There may be symptoms and signs of ischaemia or infarction in the territory of 1. the ophthalmic artery and 2. the cerebral branches of the I.C.A. Besides, direct signs of stenosis or thrombosis may be obtained where the artery is within reach of clinical examination. In the present paper only a short survey of the main symptoms and signs is intended.

1. Visual Symptoms and Signs

Retinal or optic nerve ischaemia, or both, result in monocular blindness.

Permanent blindness on the side opposite to an hemiplegia is of course highly suggestive of I.C.A. occlusion but it is an infrequent finding. It was present in 4% of the cases of the series of Castaigne, Lhermitte, Gautier and Lainée (1962). Permanent monocular blindness implies that either anatomical defects have prevented collateral supply to be established or that embolism or anterograde thrombus have blocked the distal branches (arteries to the retina or optic nerve) of the ophthalmic artery. The amazing cast of the ophthalmic artery removed at operation in Ross Russel's case (1961) shows that the anterograde thrombus may extend very far into the branches of the artery.

Transient monocular blindness is characterized by a sudden amblyopia or amaurosis in the whole or part of the visual field. There is no pain. The trouble lasts for seconds or minutes. In some patients vision reappears as if a curtain were drawn before the eye. Experience supports Fisher's (1952) remark that transient monocular blindness tends to stop when hemiplegia and/or I.C.A. occlusion has occurred.

In a few cases (e.g. Fisher, 1959) examination of the fundus has shown white bodies arriving, progressing and finally disappearing in the retinal arteries. This together with some pathological direct (McBrien, Bradley and Ashton, 1963) and indirect (Castaigne, Lhermitte and Gautier, case V, 1965) evidence, suggests that platelet emboli detached from a white mural thrombus may be responsible for a number of attacks of transient monocular blindness. However, it should be borne in mind that not in all cases is there firm evidence of platelet embolism.

Clinically, transient monocular blindness must, of course, be distinguished from transient hemianopia. Quite often by occluding successively each eye, the patient has made sure that the disturbance affects one eye only. When this evidence is lacking, it may be arduous to reach a firm diagnosis. It may be remembered that at least permanent definitive hemianopia as an isolated sign is probably very rare in I.C.A. disease (Alajouanine et al., 1965).

As previously mentioned, ophthalmodynamometry is a useful procedure to detect recent occlusions or tight unilateral stenoses. Arm-to-retina fluorescein time is also of interest, particularly as a tool for estimating some of the haemodynamic effects of I.C.A. stenoses.

2. Cerebral Symptoms and Signs

It needs here only to be mentioned that cerebral disorders due to I.C.A. disease result in one or several of the following cerebral disturbances: hemiplegia or hemiparesis, sensory deficit of cortical type, speech disorders when the dominant hemisphere is involved, lateral homonymous visual field defects. In 91 cases of atherosclerotic I.C.A. stenosis or occlusion, with permanent unilateral neurological disorders, ALAJOUANINE et al., 1965) found that motor deficit was present in 93%, speech disorders in 89% of patients with left hemisphere involvement, sensory disorders in 65%, visual field defects in 30%. The clinical picture in some patients was monosymptomatic: 9% of cases with motor deficit, 1.5% with sensory deficit, 1% with speech disorders and none with visual field defects.

Cerebral accidents due to I.C.A. atherosclerosis have three main clinical pictures: 1. sudden stroke with severe permanent neurological deficit; 2. progressive stroke; 3. transient ischaemic attacks.

Sudden severe strokes bring about permanent neurological deficit at once or in a matter of minutes. Loss of consciousness carries a very poor prognosis. In fatal cases of our own the infarct generally involves the territories of the middle cerebral and anterior choroïdal arteries and more or less that of the anterior cerebral artery. Tentorial herniation is almost always present.

Progressive strokes, complicated by seizures lasting several hours or even a few days, are by no means rare. They were present in 54% of 91 cases of ALAJOUANINE et al. (1965). Conceivably advantage might be taken of this protracted evolution to endeavour some therapeutic action (anticoagulant drugs?, fibrinolytic drugs?, surgical intervention?) but what is taking place in the artery is not really well known. It can be imagined that a thrombus goes through periods of expansion and retraction, thus being more or less obstructing; it might equally well be that successive emboli plug more and more cerebral arteries; it might also be that collateral supply undergoes significant variations; besides, it might be that the onset of the stroke results from the occlusion of the I.C.A. sinus and that the progression of neurological disorders results from the extension of the anterograde thrombus. Clearly, the riddle of this clinical picture deserves further research.

Transient ischaemic attacks obviously carry the highest interest since they allow the hope of finding the arterial lesions before the onset of irretrievable cerebral damage. The mechanisms of transient cerebral ischaemia due to atherosclerosis have for long been a subject open to conflicting views. While most authors probably agree that arterial spasm is at present out of fashion, some still support the role of falls in blood-pressure. On the other hand, some patients who have experienced transient monocular blindness have also experienced transient cerebral attacks. Therefore, there is ground to think that at least some transient ischaemic attacks are due to platelet emboli which detach from I.C.A. stenoses and crumble after having occluded an artery for a short while.

Other Symptoms and Signs: Headache is frequent in I.C.A. occlusions and may be present in tight stenoses. Patients generally complain of a dull, throbbing pain, above the eye and over the temple. *Myosis,* with or without a faint ptosis, is occasionally encountered. It is generally held that this results from involvement of the sympathetic fibers which run upwards upon the I.C.A. However infrequent, this sign of I.C.A. disease is of interest since Horner's syndrome may also be present in brain stem disease, a fact that could lead to misinterpretation. Focal or generalized *epilepsy* is rare at the onset of cerebral disorders. *Diplopia* is also rare but of interest since, like myosis, it could well lead the diagnosis to vertebro-basilar disease. The mechanism of diplopia in I.C.A. disease is not clear. Some disorder of the nutrient arteries to the oculo-motor nerves in the cavernous sinus may obviously be imagined. However, present knowledge about the arterial suply of the oculo-motor nerves in the cavernous sinus (DREYFUS, HAKIM and ADAMS, 1957) is, in our opinion, insufficient to reach firm conclusions. *Psychological disturbances* e.g. intellectual impairment, have been reported. *Cervical pain* is present in some patients at a time apparently corresponding to that of the occlusion.

Particular Clinical Pictures

1. Occlusion of both I.C.A. is not very rare, a fact accounted for by the prevalence of bilateral atherosclerotic lesions (s. above). Each occlusion may result in two successive severe accidents. On the other hand, when transient ischaemia involves successively each cerebral hemisphere or when after a first occlusion having been clinically silent the second results in bilateral neurological deficit, the clinical picture closely resembles that of stenosis or occlusion of the basilar artery. A peculiar motor and psychological deterioration has been reported (CASTAIGNE, LHERMITTE, CAMBIER and GAUTIER, 1963).

2. In some patients, occlusion of an I.C.A. may determine a very progressive neurological deficit, the clinical picture having much in common with that of a cerebral tumour.

3. Clinical Examination of the I.C.A.

Palpation of the proximal I.C.A. is not of much help, for pulsations from the common or external carotid arteries are usually present even if the I.C.A. is occluded. Moreover, palpation should always be cautious for reasons similar to those mentioned below with digital compression.

Digital *compression* was been widely in use not so long ago. It is certainly a simple clinical test: when one I.C.A. is occluded, the digital compression of the other one results in a loss of consciousness in a few seconds. However, positive or negative, the test does not replace angiography; therefore its value is limited. Moreover, it is not a safe test: admittedly, accidents are not frequent but they have occurred and were due, most probably, to the breaking off of a mural thrombus.

Auscultation of the I.C.A. is one of the most important of the modern acquisitions of clinical neurological examination. The great majority of carotid bruits are heard over the common carotid bifurcation area. Bruits heard over the eyeball and resulting from a stenosis of the carotid syphon are very rare. Not every stenosis, of course, gives rise to a bruit and bruits which do not result from a stenosis are fairly frequent. However, in a patient with cerebrovascular disease, the presence of a well defined bruit is highly suggestive of a carotid lesion. With such clinical criteria, angiography was positive in 75% of the cases in a series of RENNIE, EJRUP and McDOWELL (1964). A new and often bothering problem for neurologists is the discovery of a carotid bruit in a patient free of other symptoms or signs. Should angiography be always performed in such a condition? A satisfactory decision is not easy to reach. It might be remembered that McDOWELL and EJRUP (1966) have shown that in patients over 40 years of age, a bruit in the neck as a single sign is not particularly a harbinger of disaster at short term. Some of these bruits eventually disappear. In their present experience, the authors of this paper take into account the pressure level in the retinal arteries, since a low pressure on the side of a bruit is very suggestive of a tight stenosis and therefore should make angiography more commendable.

References

ALAJOUANINE, T., LHERMITTE, F., GAUTIER, J. C.: Accidents ischémiques cérébraux d'origine athéroscléreuse. Etude clinique et artériographique. Gaz. sanit. **36**, 507—511 (1965).

BICKERSTAFF, E. R.: Aetiology of acute hemiplegia in childhood. Brit. med. J. **2**, 5401, 82—87 (1964).

BRICE, J. G., DOWSETT, D. J., LOWE, R. D.: Haemodynamic effects of carotid artery. stenosis. Brit. med. J. **5421**, 1363—1366 (1964).

CASTAIGNE, P., LHERMITTE, F., CAMBIER, J., GAUTIER, J. C.: Obstruction bilatérale des carotides internes. Etude anatomo-pathologique d'une observation avec survie prolongée. Presse méd. **71**, 757—760 (1963).

— — GAUTIER, J. C.: Rôle des lésions artérielles dans les accidents ischémiques cérébraux de l'athérosclérose. Rev. neurol. **113**, 1, 1—32 (1965).

— — — LAINÉE, J.: Correlations cliniques et artériographiques dans 250 cas d'accidents ischémiques du cerveau d'origine athéroscléreuse. Rev. neurol. **106**, 497—501 (1962).

DREYFUS, P. M., HAKIM, S., ADAMS, R. D.: Diabetic ophthalmoplegia. Arch. Neurol. Psychiat. (Chic.) **77**, 337—349 (1957)

DUGUID, J. B.: Thrombosis as a factor in the pathogenesis of coronary atherosclerosis. J. Path. Bact. **58**, 2, 207—212 (1946).

— Thrombosis as a factor in the pathogenesis of aortic atherosclerosis. J. Path. Bact. **60**, 1, 57—61 (1948).

— Pathogenesis of atherosclerosis. Lancet **1949 II**, 6586, 925—927.

FISHER, C. M.: Transient monocular blindness associated with hemiplegia. Arch. Ophthal. **47**, 167—203 (1952).

— Occlusion of the carotid arteries. Arch. Neurol. Psychiat. (Chic.) **72**, 187—204 (1954).

— GORE, I., OKABE, N., WHITE, P. D.: Atherosclerosis of the carotid and vertebral arteries. J. Neuropath. exp. Neurol. **24**, 3, 455—476 (1965).

GAUTIER, J. C.: Genèse et évolution des thromboses dans la circulation cérébrale. Symposium International sur la Circulation Cérébrale, p. 255—260. Paris 15—16 Oct. 1965. Paris: Sandoz 1966.

GILLAND, O., HELGASSON, L., LEISSNER, P., STATTINS, S.: Arm-retina fluorescein circulation time in carotid obstruction. Acta neurol. scand. **41**, suppl. 13, part. 1, 319—333 (1965).

HULTQUIST, G. T.: Über Thrombose und Embolie der Arteria Carotis. Thèse, No. 9, Stockholm. Stockholm: G. Fisher 1942.

LENÉGRE, J., HIMBERT, J., BEAUMONT, J. L., FROMENT, R., NORMAND, J., PERRIN, A.: Athérosclérose et insuffisance coronarienne. Rapport au XXXIIIe Congrès Français de Médécine. Paris: Masson 1961.

LHERMITTE, F., GAUTIER, J. C., DEROUESNÉ, C.: Anatomopathologie et physiopathologie des sténoses carotidiennes. Rev. neurol. **115**, 641—672 (1966).

— — — GUIRAUD, B.: Ischemic accidents in the middle cerebral artery territory. A study of the causes in 122 cases. Arch. Neurol. (Chic.) **19**, 248—256 (1968).

Luessenhop, A. J.: Occlusive disease of the carotid artery. Observations on the prognosis and surgical treatment. J. Neurosurg. 16, 705—730 (1959).

McDowell, F., Ejrup, B.: Arterial bruits in cerebrovascular disease. A follow-up study. Neurology (Minneap.) 16, 1127—1129 (1966).

Mitchell, Jr., A., Schwartz, C. J.: Arterial disease. Oxford: Blackwell Scientific Publications 1965.

Pitner, S. E.: Carotid thrombosis due to intraoral trauma. An unusual complication of a common childhood accident. New Engl. J. Med. 274, 14, 764—768 (1966).

Poole, J. C. F., French, J. E.: Thrombosis.. J. Atheroscler. Res. 1, 3, 251—282 (1961).

Rennie, L., Ejrup, B., McDowell, F.: Arterial bruits in cerebrovascular disease. Neurology (Minneap.) 14, 751—756 (1964).

Ross Russel (W. R.): Observations on the retinal blood vessels in monocular blindness. Lancet 1961 II, 7218, 1422—1428.

Shillito Jr., J.: Carotid arteritis: A cause of hemiplegia in childhood. J. Neurosurg. 31, 6, 540—551 (1964).

Tindall, G. T., Odom, G. L., Cupp Jr., H. B., Dillon, M. L.: Studies on carotid artery flow and pressure. Observations in patients during graded occlusion of proximal carotid artery. J. Neurosurg. 19, 11, 917—923 (1962).

Yates, P. O., Hutchinson, E. C.: Cerebral infarction. The Role of stenosis of the extracranial cerebral arteries. Med. Res. Counc. Spec. Rep. Sér. No. 300. London: Her Majesty's Stationery Office 1961.

Microangiographic Study of Ischemic Cortical Lesions (Outside the Supply-Territory) Following Carotid Thrombosis*

H. Gastaut, G. Salamon, and M. Toga

The existence of necrotic lesions outside a vascular supply territory or at its "border-line" is well known, but exact evidence of such observations is still lacking.

For this reason we think it interesting to report two anatomico-clinical observations in which we were not only able to make a complete examination of the patients, but also were able to study the arterial cerebral network by micro-angiography. This study of the arterial network forms a basis for interesting discussion of the electro-clinical correlations.

Case No. 1: M. Bon, Leon, 65 years old, had suffered a leftsided hemiplegia 7 years ago from which he had partly recovered. After 5 years—i.e. 2 years before he was taken into hospital—he had a more severe new leftsided hemiplegia with a smaller degree of recovery. There remained a very marked neurological deficit in his left arm.

Eight days prior to his admission he had clonic fits in the left leg and also some speech disturbance. An epileptic status in the leg developed. Since his admission the fits in the leg persisted in spite of intensive treatment and the patient died after 4 days.

The EEG (Fig. 1) showed elements of low amplitude and complexity or an alpha activity which reacted little and was present over both hemispheres. It was partly masked by slower rhythms of the theta or delta band. Most important seemed to be a much lower amplitude of all activity in the right hemisphere. Our conclusion was that of a general very severe cerebral lesion more marked, however, on the right where the electrical activity was depressed. Angiography of the right carotid showed a total thrombosis in its cervical segment. By cervical angiography with subtraction (Fig. 2), a certain reflux into the carotid syphon from the ophthalmic artery was seen. A slow perfusion with a fine solution of barium after early autopsy of the brain was made for the microangiographic study. The brain was fixed and frontal slices of 1 cm thickness made. Exposure on metallographic films showed the cortical fine arterial network, as well as that of the white substance and the perforating arteries (Figs. 3, 4, 5).

The sections were placed exactly parallel to the commissures. Microangiographically, there was no change in the vascular network of the left hemisphere. On the right, the supply territories of the three great arteries were sufficiently injected to show a marked zone of ischemia at the borderline between the anterior and middle cerebral arteries. Here the cortical network was not visible and the arteries of the white substance were only very slightly affected. The vasculature not only "lacked" visibility in the upper third of the Rolondic zone, but also in a strip from front to back along the frontier of the second and third frontal convolutions (Fig. 17). The basal ganglia, however, were not involved.

* From the Centre Hôpitalier et Universitaire de Marseille, Service d'Exploration Fonctionelle du Système Nerveux, Hôpital de la Timone, Bd. Jean-Moulin, Marseille.

Anatomical Examination (No. 176/68)

The macroscopical examination of the frontal sections (Figs. 3, 4, 5 and 6) showed on the right an older softening in a stage of greater or lesser organisation unequal in degree in its various parts. Frontally, the infarct was triangular at its cortical base and reached down almost to the lateral ventricle (A 3). The following sections show

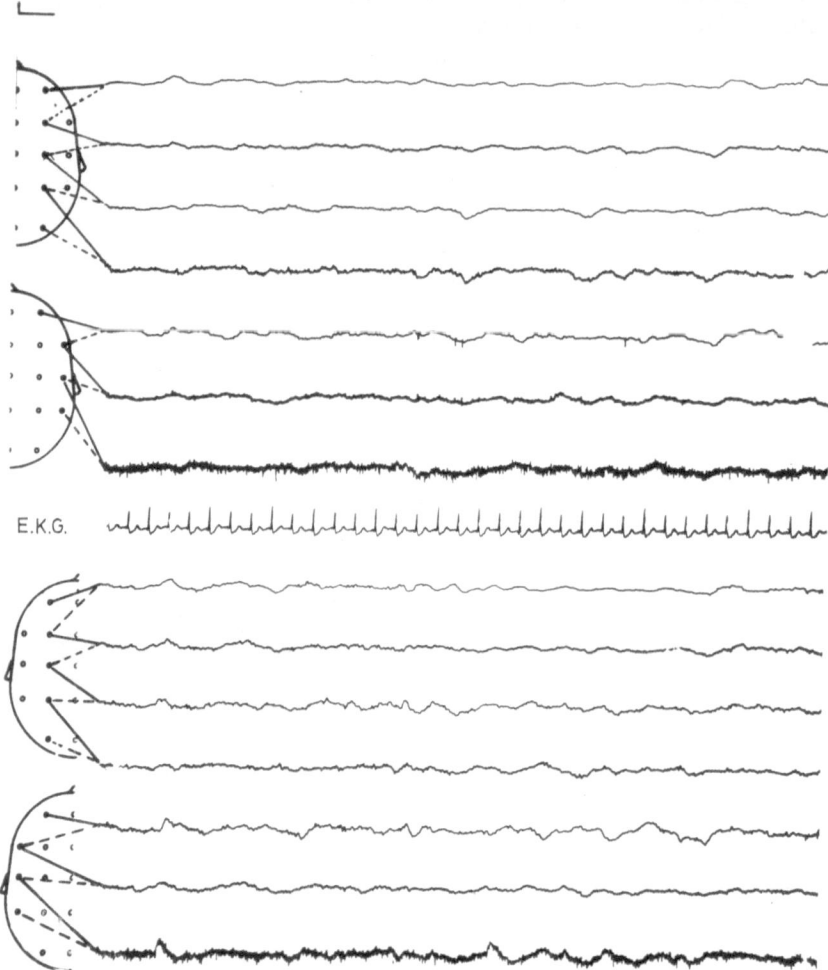

Fig. 1. First observation. EEG, Patient Bon. (see text)

the lesion on F. 1 and 2 and, still more occipitally, the extension into the parietal convolutions in P 1 and 2. The occipital lobe itself was not involved. No lesion was visible in the left hemisphere.

Histological Examination

There was very severe demyelination on the right side corresponding to the macroscopical lesions. Moreover, an extensive formation of cysts was visible which was traversed by delicate fibrillary tracts joining the intact cortical region and the centrum semiovale. In the zone of destruction ganglion cells had been lost and scavenger cells and protoplasmic and fibrillary astrocytes were visible. Below the zone of softening there was a patchy loss of neurones, a moderate degree of astrocytosis and an infiltration of the subependymal zone with corpora amylacea. All intracerebral arteries showed a thickening of the walls and a sclero-hyalinotic change.

Case No. 2: N. Cas. Jques, François who for 3 months showed first a right hand paresis, then a palsy of the left arm. He had a bilateral paresis of the legs and memory defect; on admission he showed signs of diffuse pyramidal deficit and also had disturbance of speech. Two days after his admission he had clonic fits in his right arm. He died on the 5th day.

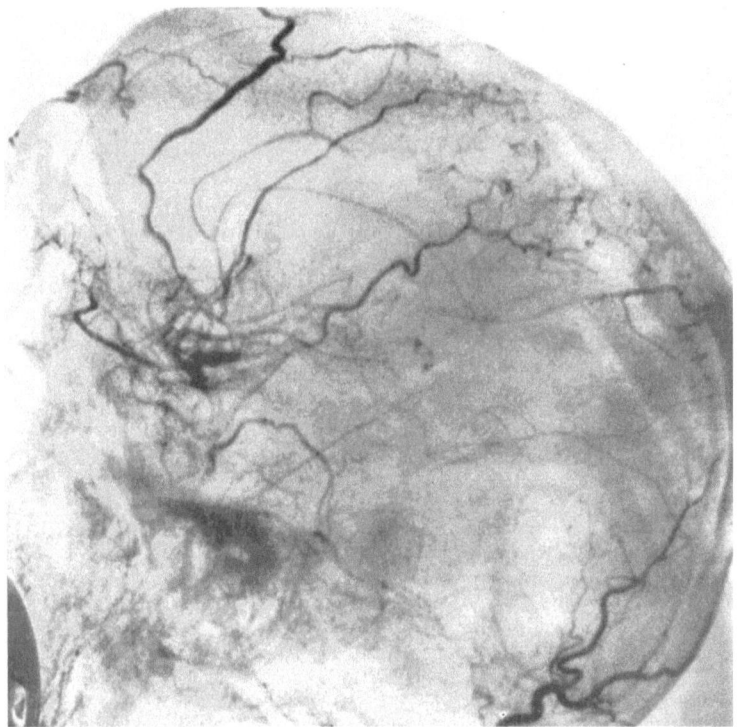

Fig. 2. First observation. Patient Bon. Right carotid angiography. Thrombosis of the internal carotid with slight filling of the syphon by the ophthalmic artery

The EEG (Fig. 7) was similar, if not identical, with that of the previous case: it showed a basic complex rhythm or an alpha activity which did not react to outer stimuli. Superimposed were slow waves of the theta/delta band. There was an asymmetry, the right hemisphere being globally depressed in its activity. In summary: signs of a diffuse cerebral lesion occurred predominantly on the right where there was depression of activity.

Right sided carotid angiography (Fig. 8) showed complete occlusion of the internal carotid artery with a certain filling of the syphon by a very voluminous ophthalmic artery. The meningeal and temporal superficial arteries were very well developed.

Left carotid angiography (Fig. 9) also showed occlusion of the internal carotid. Supplementary anastomoses also filled the syphon and some faint vascular branches.

With these findings of a double sided carotid occlusion, it was decided to make a vertebral angiography. This turned out to be normal. The vertebro-basilar system apparently did not help in supplying the two carotid systems (Fig. 10).

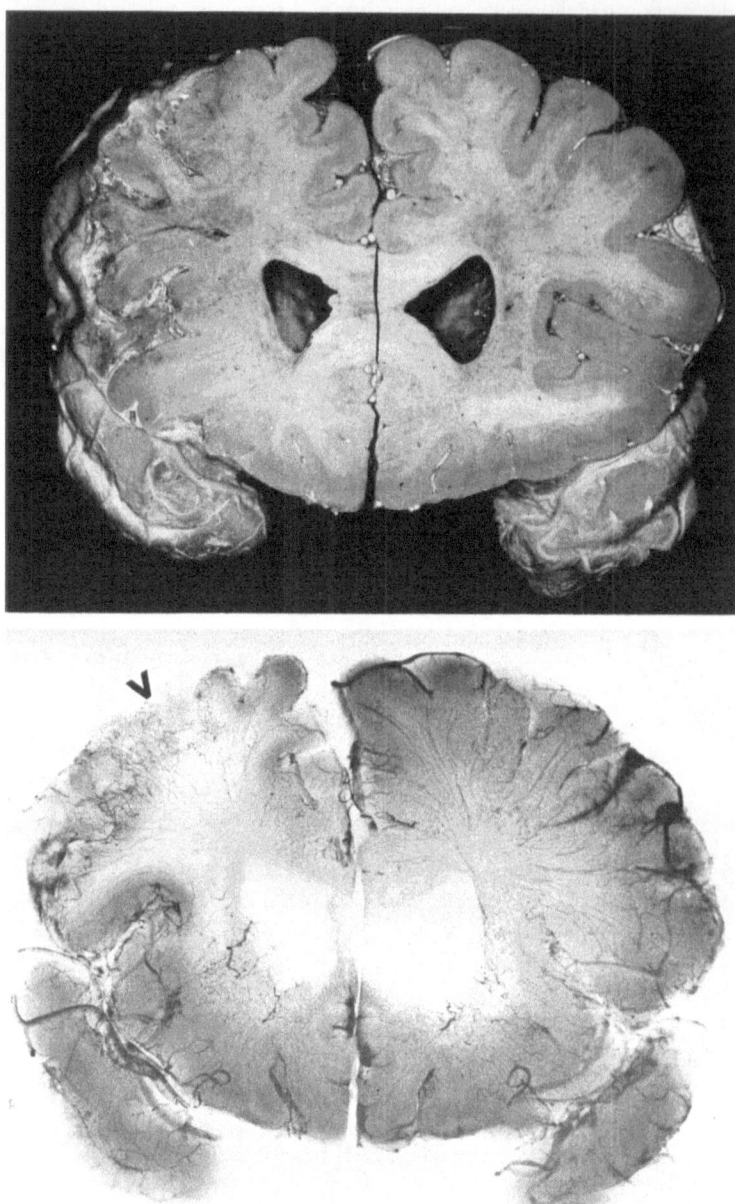

Fig. 3. First observation. Patient Bon. Vertical section in front of the caudate nucleus. Comparison of the macroscopic and microangiographic picture. At the frontier of the F_1 and F_2 convolutions a lesion is visible, which in microangiography corresponds to the "frontier zone" between the anterior and middle cerebral artery territories

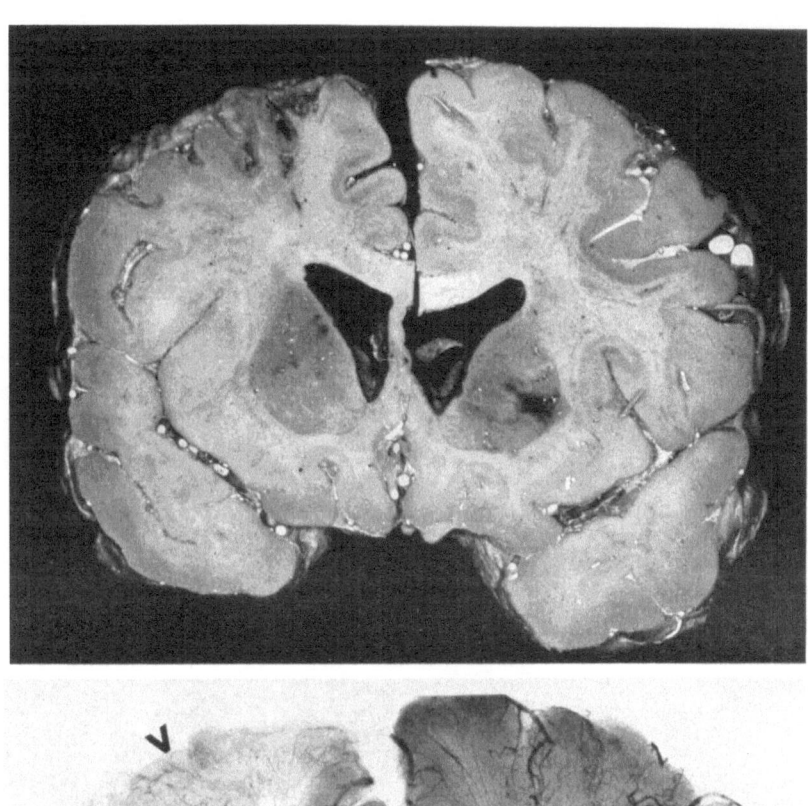

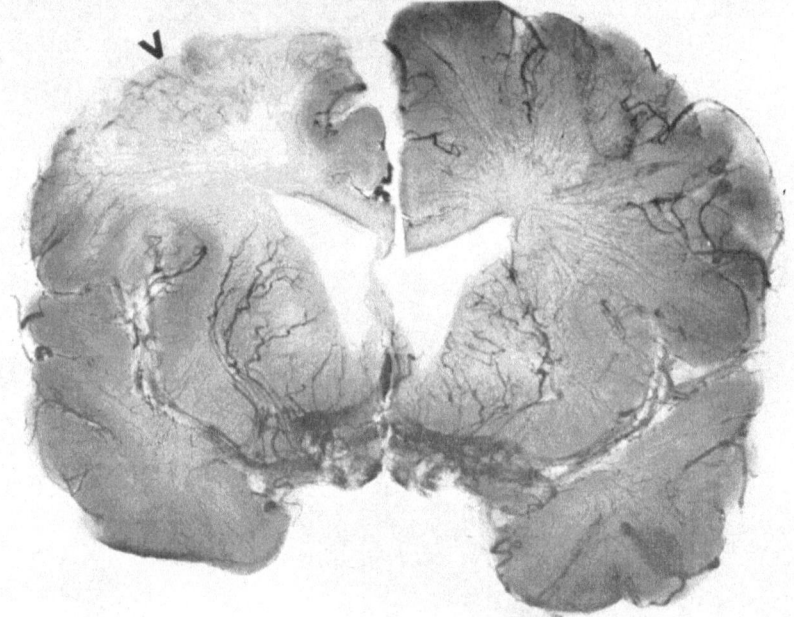

Fig. 4. First observation. Patient Bon. Vertical section through the nucleus caudatus. In the zone of transit between the territories of the anterior and middle cerebral arteries a softening is seen with disappearance of the circulatory pattern

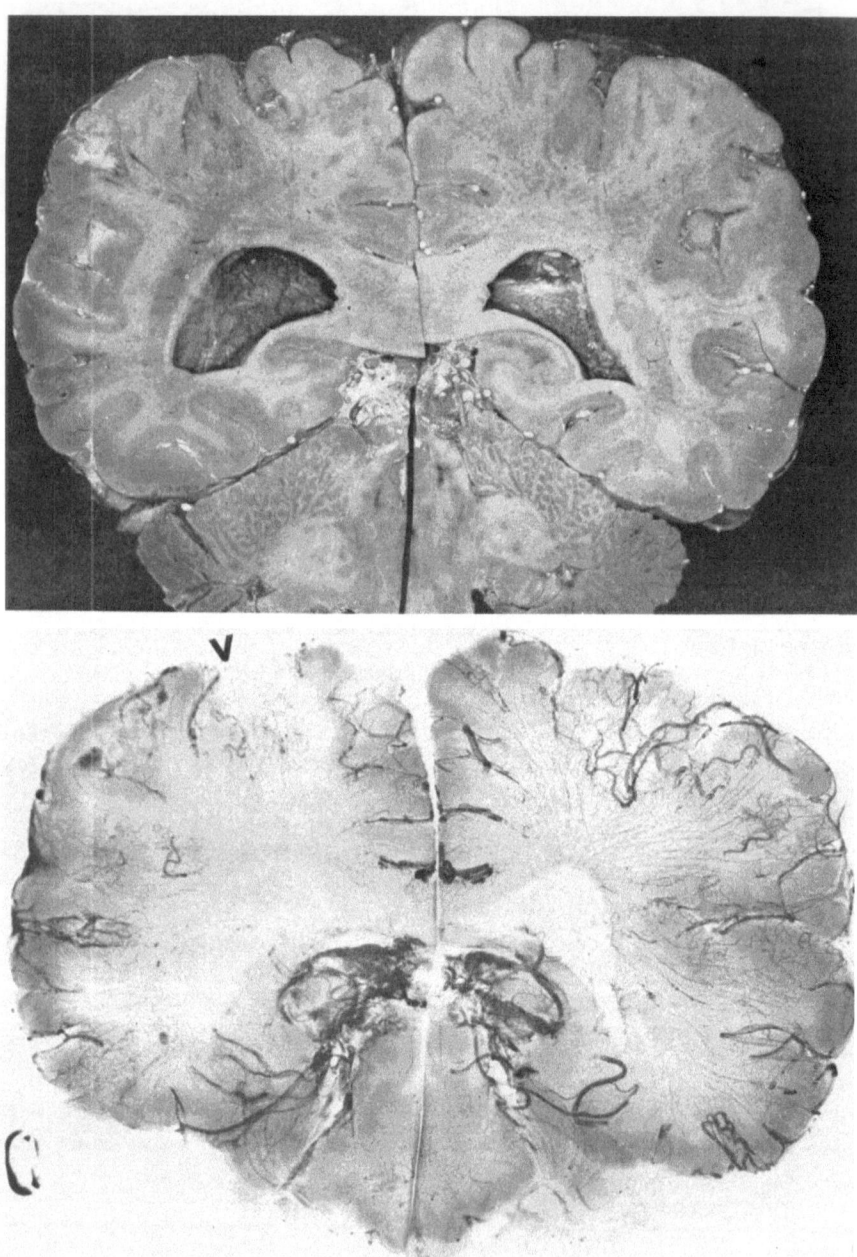

Fig. 5. First observation. Patient Bon. Frontal section through the trigone. The topography of the lesions corresponds exactly to those of the Fig. 3 and 4. The rest of the circulation is normal

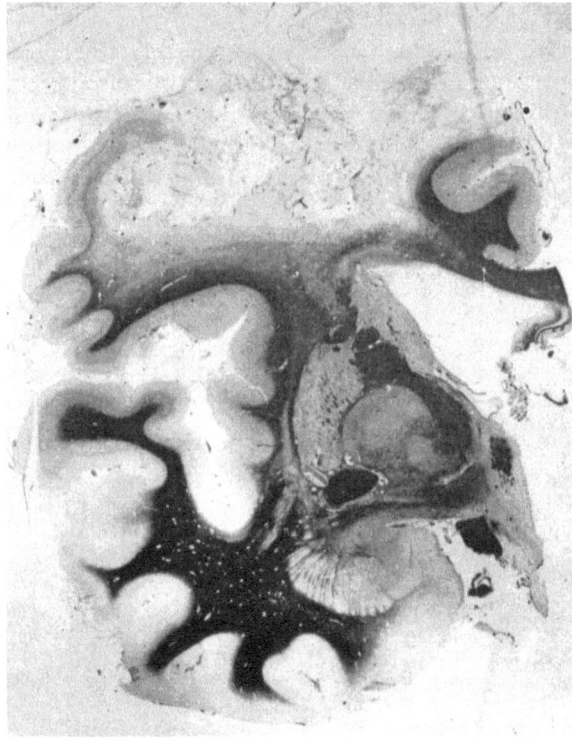

Fig. 6. First observation. Patient Bon. Cystic softening destroying the F_2 and F_3 convolutions, whereas F_1, the cingulate gyrus and the island of Reil are preserved. Myelin stain

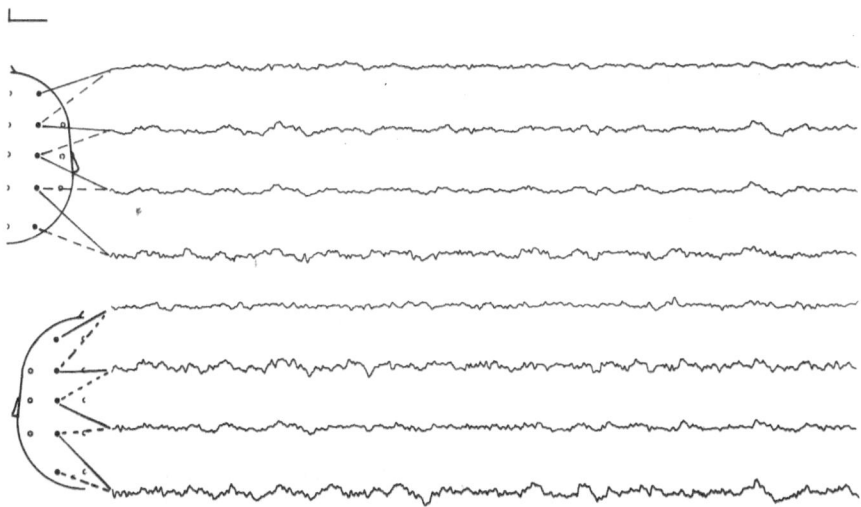

Fig. 7. Second observation. Patient Cas., EEG (see text)

Microangiographic Study

After early autopsy, the injection was made by the same technique as in the previous case. The study of the serial sections (Figs. 12, 13, 14) showed the location and size of the lesion. The vertebro-basilar system was entirely normal. On the left, the territory of the middle and anterior cerebral arteries could be equally regarded as normal. However, as in the previous case, the frontier zone between these two great

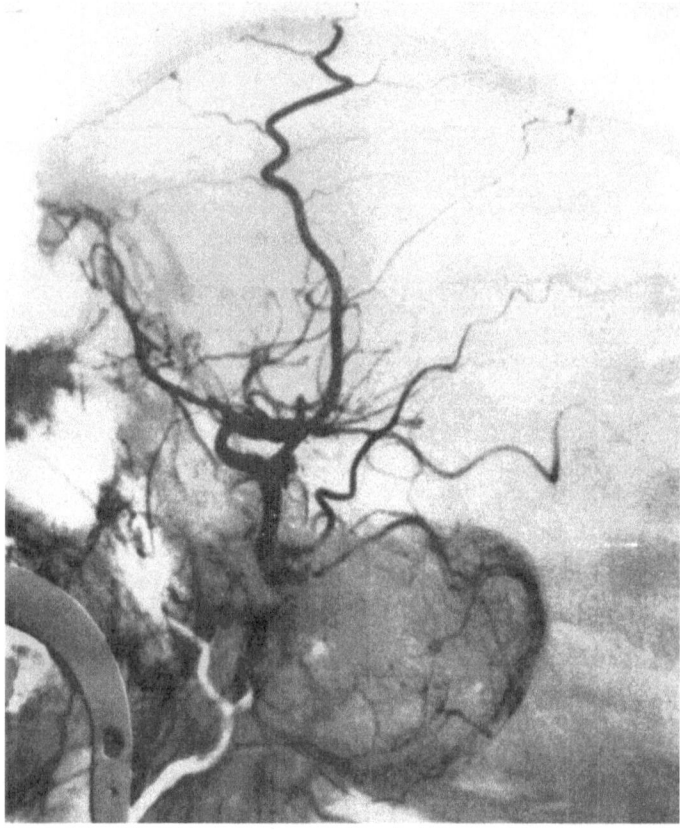

Fig. 8. Second observation. Pat. Cas. Right carotid angiography. Thrombosis of the right carotid with abundant collateral circulation by the ophthalmic artery

arterial territories showed a loss of the cortical vasculature and its branching into the white substance.

These lesions were located from frontally to occipitally, according to the same "strip" pattern (Fig. 17) as in the previous case.

On the right side, the lesions were similar, yet far more extensive. The cortical ischemia of the terminal parts of the sylvian territory and that of the anterior cerebral artery were far more severely changed. Also, part of the white substance had disappeared, a lesion which reached to the caudatum. There was also an ischemic area at the deep frontier zone between the sylvian cortical and the deep basal ganglia branches of the same artery.

Anatomical Examination

During macroscopic examination of the fronto-occipital series of slices, an infarct was found in the right hemisphere which was surrounded by an area of petechiae which was located in the frontal lobe in the white substance of F 3 and the centrum ovale.

The infarct had destroyed the second and third convolutions, the caudate nucleus on the A 2 section and the upper part of the putamen as well as the upper lip

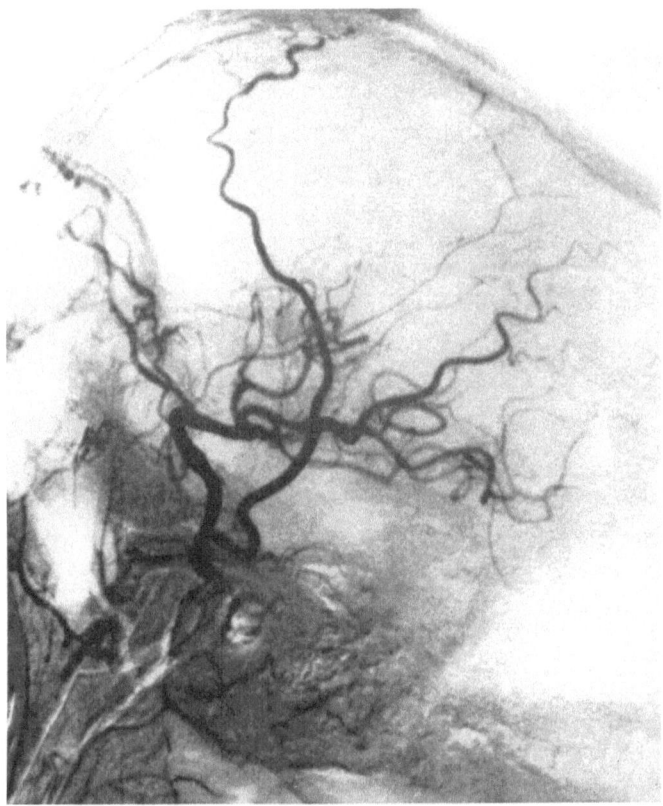

Fig. 9. Second observation. Patient Cas. Left carotid angiography with thrombosis of the left internal carotid. The syphon of the carotid and some of its major branches are filled by the ophthalmic collateral circulation

of the insula on the A 1 section. More distally on the P 1 section the softening had involved the centrum semiovale, the ventricular angle and the white matter of the parietal convolutions. Even more distally, the sections showed the infarction of the parieto-occipital convolutions and of the deep nuclei (Figs. 12—16).

In the left hemisphere, the corpus callosum was partially destroyed at its beginning near the centrum semiovale by a very well limited infarct zone. This destruction zone was wider in the A 1 section and comprised here F 1 and 2, whereas on the P 3 section it had destroyed the upper parietal convolution. The occipital pole again had been spared.

Histological Examination

Disintegration of the myeline sheaths on both sides from frontally to occipitally, corresponding to the macroscopical observations. With aniline stains, there was cystic degeneration with formation of cavities traversed by fine vascular glial filaments. There

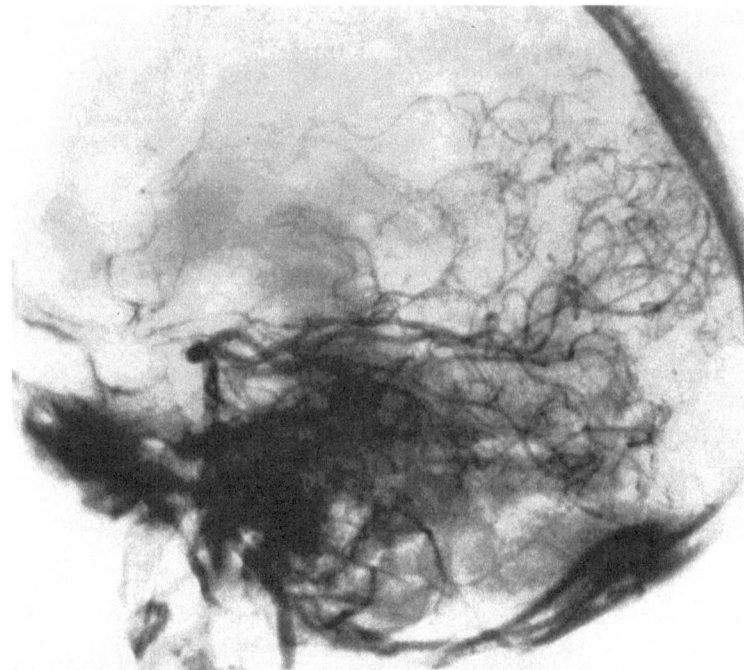

Fig. 10. Second observation. Patient Cas. Vertebral angiography. The posterior communicating arteries have failed to function as a collateral system. The vertebro-basilar system seems normal

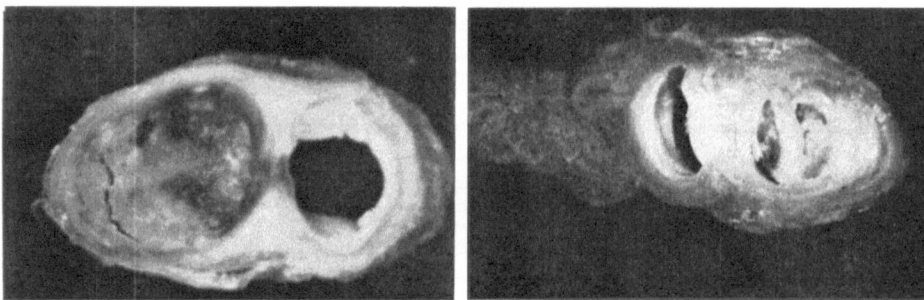

Fig. 11. Second observation. Transverse section of the two bifurcations shows the occlusion of the internal carotid artery in both

were numerous histiocytes, macrophages and scavenger cells partly laden with hemosiderin. In the region of the cystic degeneration, there were multiple small polygonal necrotic foci with their axes parallel with the surface. Here only the molecular layer was intact. The parenchymal cells were lost and many fatty scavenger cells were present.

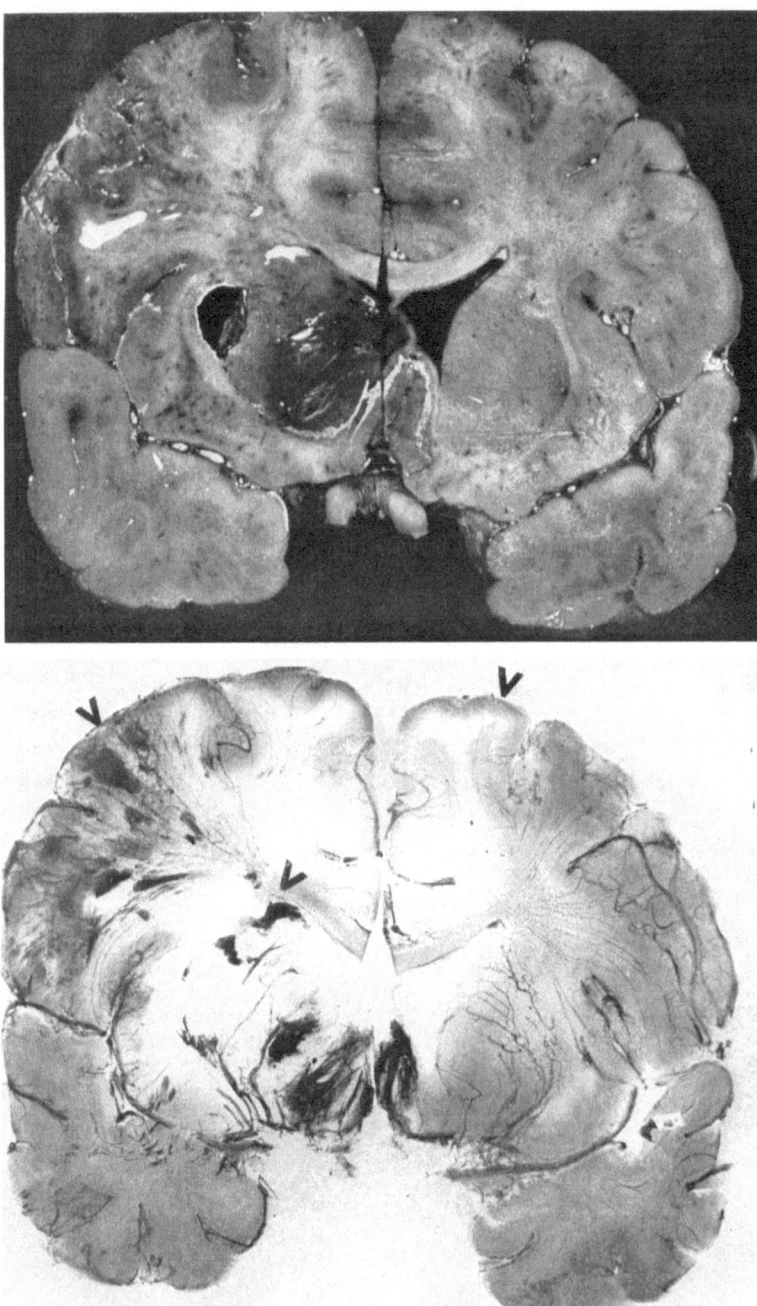

Fig. 12. Second observation. Patient Cas. Vertical section showing the macroscopical and microangiographic aspect. On the right, one sees the softenings and the absent circulation in F 2/3 i.e. in the frontier zones between the anterior and middle cerebral arteries. Moreover, a similar lesion exists between the superficial and deep system of the Sylvian artery. On the left, one finds a similar but less extensive lesion

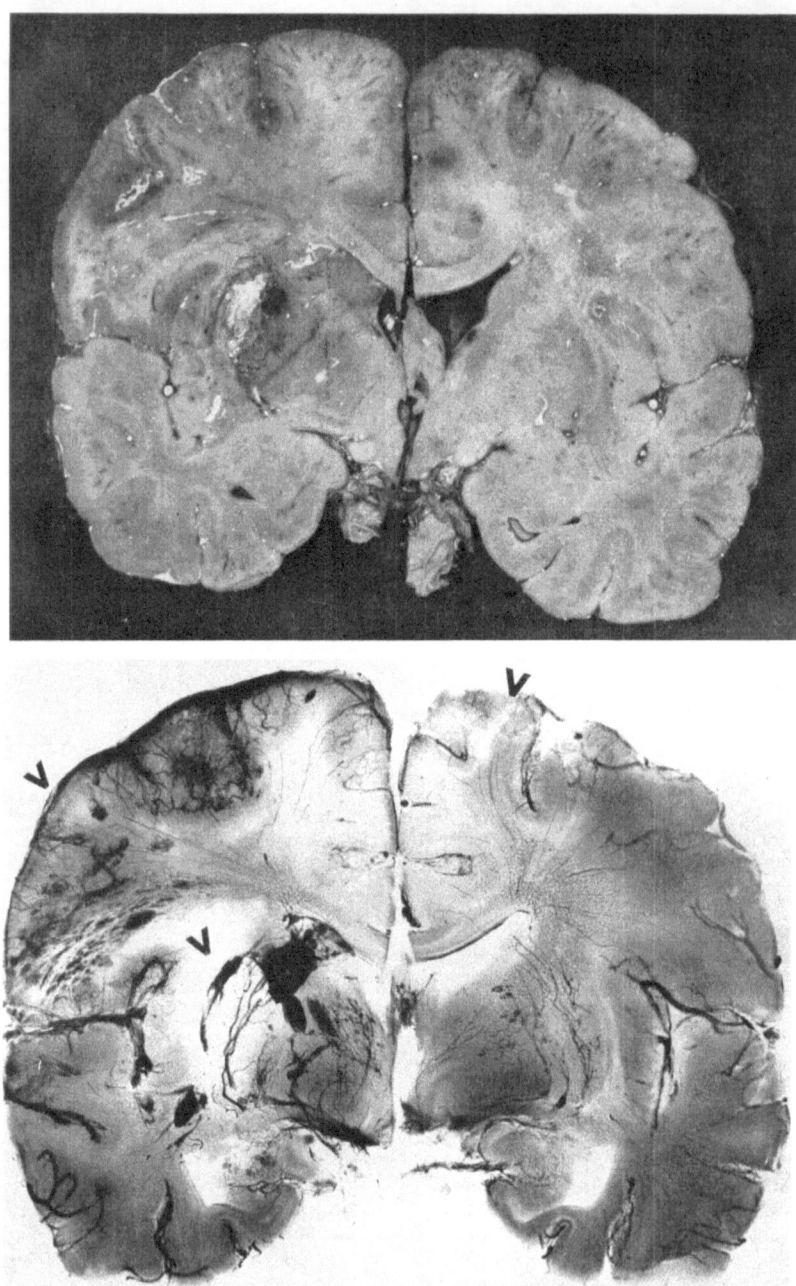

Fig. 13. Second observation. Patient Cas. Vertical section through the anterior plane of the thalamus. The lesions resemble that of Fig. 12

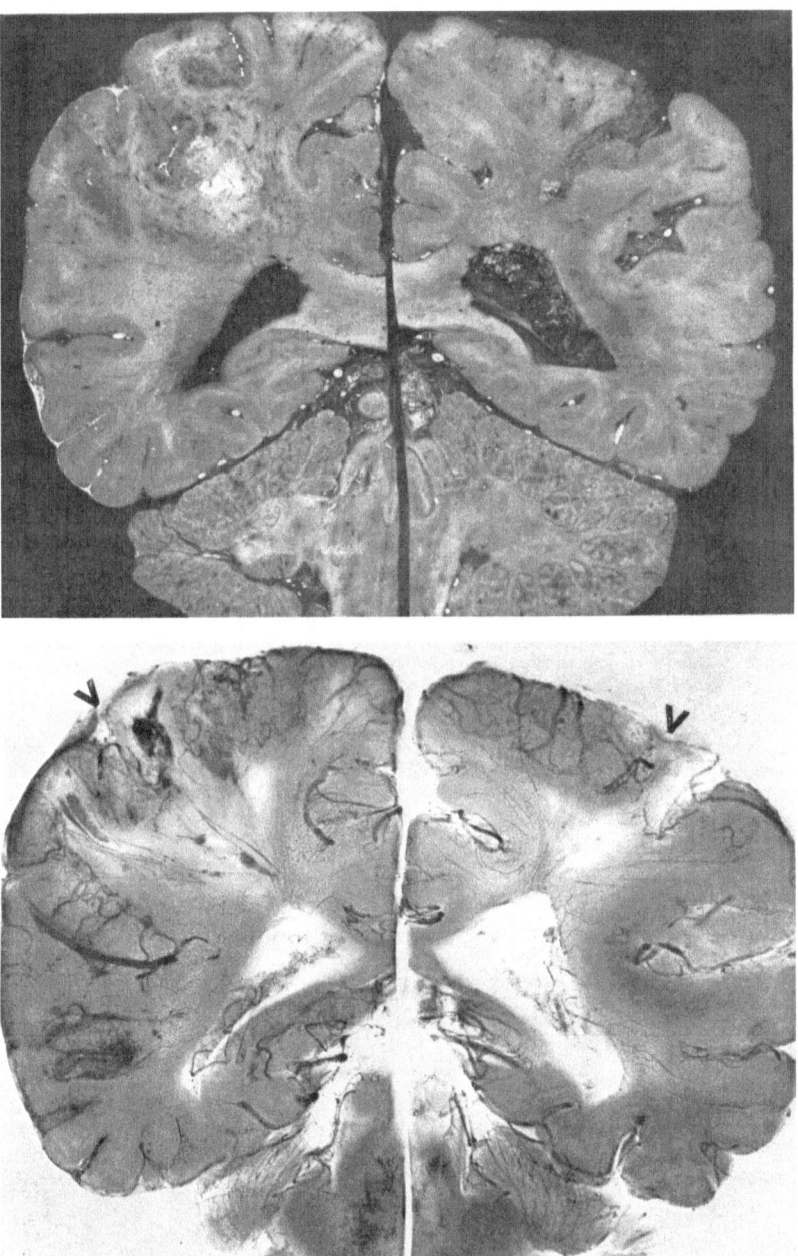

Fig. 14. Second observation. Patient Cas. Vertical section through the trigone. On both sides lesions of the "frontier-line" type are visible

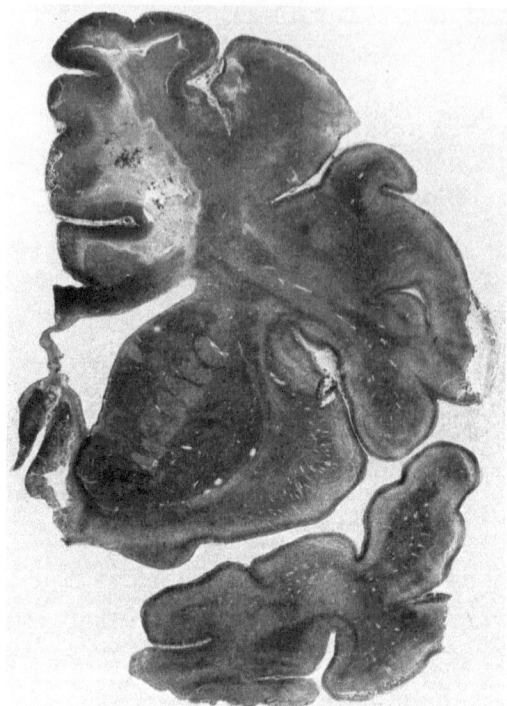

Fig. 15. Second observation. Patient Cas. The white substance of the cingulate gyrus and of F 1, the cortex of F 2 and the beginning of the corpus callosum are destroyed by softening

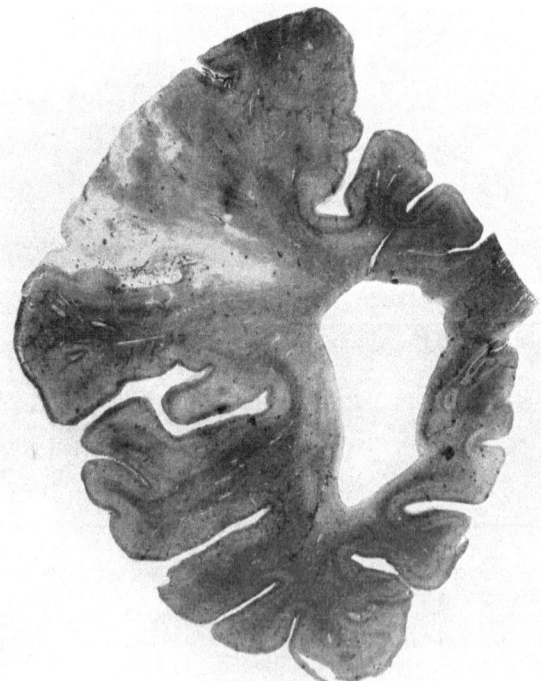

Fig. 16. Second observation. Patient Cas. On the leftside destruction is limited to the convolutions P 1 and P 2

At a distance from these necrotic foci, there was only a loss of neurones or signs of a chronic lesion (retraction, hyperchromasia, pyknosis of the nucleus etc.). In the cerebellum, there appeared necrosis of one layer only i.e. of the granular cells, whereas the Purkinje cells were relatively spared. The vessels seemed to have a thickened wall. The carotid arteries showed a thrombosis in various segments (Fig. 11).

Comments. Electro-clinical Correlations

These cases show the variety of changes in cases of carotid thrombosis, with which we are so familiar nowadays. Apart from the latent forms, which are only detected by angiography, there are the severe cases often starting with fits. The development may proceed over years—as in our first case—or start much more rapidly as in our second. Finally, there are cases which begin with paresis but then develop local fits in the same region: the first of our patients started with an epileptic status in the leg, the second with clonic fits in the right arm. There is then no typical symptomatology in the case of a carotid thrombosis.

The EEG may be normal or show focal slow waves in the sylvian territory (expressing sylvian vascular insufficiency); it may also present slow waves in the temporoparieto-occipital region which are sometimes associated with periodic intermittent sharp waves (which expresses an ischemia in an "extraterritorial" region i.e. a frontier zone). It may even show diffuse lesions in one hemisphere or in both hemispheres, in which case the one with the thrombosis shows a depression in activity.

Angiography

This is the most reliable study in the diagnosis of a carotid thrombosis. Usually one finds the occlusion of the internal carotid artery at the bifurcation and, at the same time, a supplementary circulation by the anterior anastomotic artery of the circle of Willis or, finally, the maxillo-ophthalmic anastomosis. On the other hand, angiography does not give information about the flow beyond the occlusion. One has to keep in mind that the lack of opacity of a distal segment beyond an occlusion does not indicate an entire lack of flow in this "deprived" segment.

When a circulation is slowed there is no angiographic possibility of making this visible. Thus may be explained the discrepancy between angiographic and microangiographic findings. The microangiographic preparation showed that regions were opacified by *postmortal* injection in which on angiography of the carotid or vertebrals, there was apparently a complete loss of flow. These patterns of a supplementary circulation, which may escape angiography, may then explain in part the discrepancies between the EEG findings, the clinical course and the anatomical observations.

Microangiography

The injection of a radiopaque medium into the cerebral vasculature of patients who have died from cerebro-vascular insufficiency seems to us the most logical method for studying any slowing of the circulation or complete ischemia. Mainly by a global injection will we be able to study in a more detailed way the vascular network.

Only by microangiography were we able to show, in these two cases of thrombosis of the carotid arteries, the lesions outside the supply territory. Although the first case was a unilateral and the second a bilateral thrombosis, microangiography showed identical lesions at the frontier zone of the anterior and middle cerebral arteries (Fig. 17).

However, we may ask ourselves whether the lack of an intracerebral vascular network really does allow for the conclusion of ischemia and necrosis if serial sections have not actually been made at autopsy.

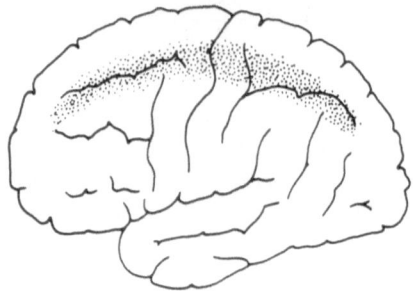

Fig. 17. Schematic drawing of the vascular lesions in our two cases

Anatomy

The unilateral lesions of our first case and the bilateral of the second have an identical pattern. They correspond to a "bandlike" zone from frontally to occipitally which had undergone softening and cyst formation.

On the vertico-frontal sections this zone seemed to have a cortical basis and to correspond to the "frontier" between the anterior and middle cerebral arteries.

Apart from the macroscopical necroses, microscopically the brain tissue was not unchanged. There were microscopical focal tissue lesions, and a diffuse atrophy with loss of neurons and gliosis as the consequence of vascular insufficiency. Moreover, a sclerosis and hyalinization of the arterial vessel walls was visible. In conclusion, we want to emphasize the topographical identity of the ischemic and the microangiographic lesions.

Conclusions

Thromboses of the carotid arteries can show either no anatomical lesions or ischemic changes in or outside the supply territory.

1. A lack of anatomical lesions is seen (Fig. 18a) when a thrombosis develops slowly and when the circle of Willis supplies a sufficient auxiliary circulation.

2. On the other hand (Fig. 18b), an interruption of carotid flow, for example by ligature in surgical cases, will lead to extensive destruction if the circle of Willis does not function sufficiently. Therefore, one sees not so rarely superficial or deep infarcts in the sylvian territory in the case of carotid thrombosis. Undoubtedly, the thrombosis will extend into the middle cerebral artery in such cases. In every case the infarct lies in the supply territory.

3. Finally, it seems necessary to stress the systematic character of the "extra-territorial" (frontier-line) lesions which may follow carotid thrombosis (Fig. 18c). The evocation of this topographical vascular concept —"extra-territorially" situated— seems to us to be frequently justified and of great interest, yet it will become difficult

to diagnose without angiographic, microangiographic and morphological studies of the carotids in the neck region. It seems highly interesting that, in the case of a proximal carotid thrombosis in which a good supplementary circulation may develop, there still may arise a systematic cortical lesion in the frontier zone between the anterior and middle cerebral supply territories. That this lesion will be probably frequent and

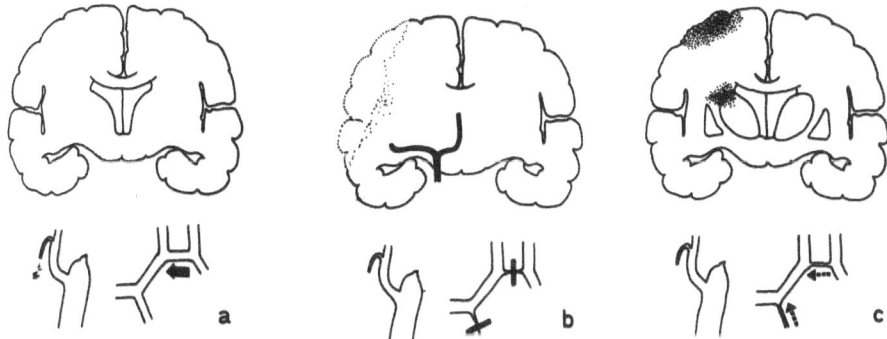

Fig. 18. a Absence of morphological lesions. b Lesions of a vascular territory. Very rapid development of thrombosis without functioning of circle of Willis. Extension of thrombosis into the Sylvian system. c Ischemic lesions of "extraterritorial" (frontier zone) type. Carotid thrombosis. Collateral supply system functionally insufficient

important may be deduced from the findings in our two cases where we could make an anatomical/angiographical/microangiographical/electrophysiological and clinical correlation of the findings.

Note

A complete bibliography is found in the following two publications:

Cerebral circulation. Progress in brain research, Vol. 1, No. 30. Elsevier 1968.

Espagno, J.: La circulation cérébrale. Du morphologique au fonctionnel normal ou pathologique, 1. Vol. (Rapport au Congrès de Neurochirurgie de Langue Française, Nancy, Juin 1969). Paris: Masson et Cie. 1969.

The Recurrent Artery of Heubner

A Morphological Study of the Blood Supply
of the Rostral Basal Ganglia in Normal and Pathological Conditions*

M. Kribs** and P. Kleihues

I. Introduction

In his monograph "Die luetischen Erkrankungen der Hirnarterien", O. Heubner in his protocol No. 11 dated September 13, 1872, described "a small but relatively well developed branch running from the proximal part of the anterior cerebral artery directly to the substantia perforata anterior where it penetrates into the brain". He was the first to demonstrate by means of dye injections that this rather inconspicuous artery belonged to the system of the penetrating arteries supplying the basal ganglia.

Many of the papers concerning the origin, course and supply area of the recurrent artery of Heubner[1] have been contradictory (Abbie, Aitken, Beevor, Böhne, Lazorthes, Kaplan et al., Almeida). The present study involved the investigation of 177 human brains. Its purposes are twofold: 1. to determine more precisely the origin, course, and supply area and 2. to relate clinical and pathological findings in selected cases of isolated infarctions to the supply area of the recurrent artery of Heubner.

II. Historical Aspects and Nomenclature

Heubner first mentioned the recurrent artery in 1872 during his systematic investigations on the supply areas of the cerebral arteries, but as late as 1874 he did not name the vessel. Duret in 1874 mentioned only multiple arteries and described them as *artères striées antérieures perforantes de la cérébrale antérieure*. Böhne also pointed out that there are frequently multiple branches. Those which supply the head and the medial third of the nucleus caudatus and related segments of the internal capsule he named *arteriae corporis striati anteriores*. Lazorthes used different synonyms. Because of its unusual length he called it l. c. *Arteria centralis longa* but, on the other hand, also proposed the names *artère télencéphalique* (Lazorthes et al., 1956) and *arteriae striatae anteriores mediales* (Lazorthes, 1961).

---

* From the Max-Planck-Institut für Hirnforschung, Abteilung für Allgemeine Neurologie, Köln-Merheim, Germany (Director: Prof. K. J. Zülch).

** Part of this work was submitted as doctoral thesis to the Faculty of Medicine, University of Köln.

[1] Abbreviations: r. a. H (recurrent artery of Heubner) a. c. a. (anterior communicating artery).

The name most commonly found in the literature is HEUBNER's *artery* or *arteria* HEUBNERI. This was first used by AITKEN in 1909. He again described multiple small branches of the anterior cerebral artery which he called the *system of* HEUBNER arguing that these, in spite of their anatomical variations, irrigate a constant area of the basal ganglia. Further synonyms which have been used occasionally are *arteriae striaticae mediales* (KAPLAN *et al.*, 1954, 1966) as well as *recurrent lenticulostriate artery* (TAVERAS and WOOD, 1964).

The term recurrent was introduced by SHELLSHEAR (1920/21) because of the characteristic retrograde course along the main trunk of the anterior cerebral artery.

The name recurrent artery of HEUBNER *(A. recurrens* HEUBNER*)* corresponds to the historical as well as to the anatomical characteristics of this branch of the anterior cerebral artery. It is well known in the literature and will be used exclusively in this study.

III. Origin and Extracerebral Course

Materials and Methods: Investigations were carried out in 177 human brains, most of which were fixed in formalin. For the description of the origin of the r.a.H., the anterior communicating artery (a.c.a.) was used as a reference point. In the case of the presence of more than one a.c.a., the results refer to the one which is most frontally located. If the r.a.H. had a multiple

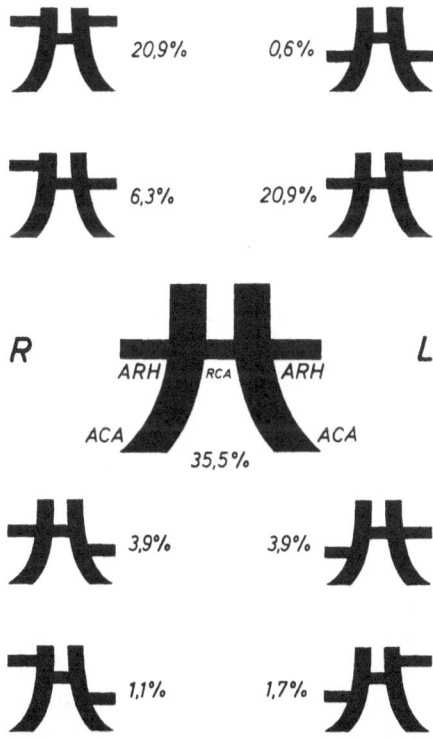

Fig. 1. The frequency of variations of the origin of the recurrent artery of HEUBNER from the anterior cerebral artery. In 5% of the 177 brains examined the recurrent artery of HEUBNER did not arise from the anterior cerebral artery (see text). *R* right, *L* left, *ARH* A. recurrens HEUBNER, *ACA* A. cerebri anterior, *RCA* R. comunicans anterior

origin, this was only classified as being proximal or distal if all branches left the anterior cerebral artery proximally or distally of the a.c.a.

Origin of the Recurrent Artery of Heubner

In 168 of 177 brains (95%) a bilateral offspring of the r.a.H. from the anterior cerebral artery was observed. The variations of the points of origin and their frequencies are summarized in Fig. 1. In more than half of the cases the r.a.H. originated bilaterally at the same level as the a.c.a. or more distally. On the other hand, a bilateral proximal origin was only observed in one case. The distance of the origin of the r.a.H. from the a.c.a. normally did not exceed 5 mm in either direction. In 8 brains the r.a.H. was not found (in 7 cases unilaterally and in 1 case bilaterally).

In 1 of these 8 cases it could be demonstrated by indian ink injection that the intracerebral area normally supplied by the r.a.H. was irrigated by the most proximal lenticulostriate branch of the middle cerebral artery. In the remainder of cases no further analysis by means of dye injection was possible because the brains were already fixed. By following the extracerebral course of the arteries penetrating the base of the brain, the possibility of the r.a.H. arising from the carotid artery or the anterior choroidal artery could be excluded.

In 21 cases there was a double origin of the r.a.H. (10 times on the left, 6 times on the right and in 5 cases bilaterally). A triple origin was only observed in a single case.

Extracerebral Course

The extracerebral segment of the r.a.H. demonstrated striking individual variations, the total length being between 1.5 and 3 cm. The major reason for variation in length was not due to the different points of origin but rather to coiling of the r.a.H. Shortly after its origin, the r.a.H. runs close to the anterior cerebral artery, often being adherent. The position is either lateral or superior to the anterior cerebral artery. At the trifurcation of the carotid artery it changes the retrograde course and now follows the direction of the middle cerebral artery.

Shortly after its offspring from the anterior cerebral artery, fine branches are observed leading to the tractus olfactorius and the base of the frontal lobe (Fig. 3). These branches vary considerably in size and number. Often they are only visible after injection of a dye. It must be pointed out that these branches can also originate separately from the anterior cerebral artery. In some cases we observed a common origin with the r.a.H.

The major branching of the r.a.H. normally occurs in the last third of the extracerebral segment. In many cases, however, a more proximal branching could be observed but only very rarely in the first third of the extracerebral segment. Usually the r.a.H. branches into 2 to 3 vessels of different sizes. These may further divide before they penetrate the base of the brain in the lateral part of the substantia perforata anterior. The largest of these extracerebral terminal branches normally enters the most lateral portion of the substantia perforata anterior. Extracerebral anastomoses between these and the penetrating branches of the middle cerebral artery have not been observed.

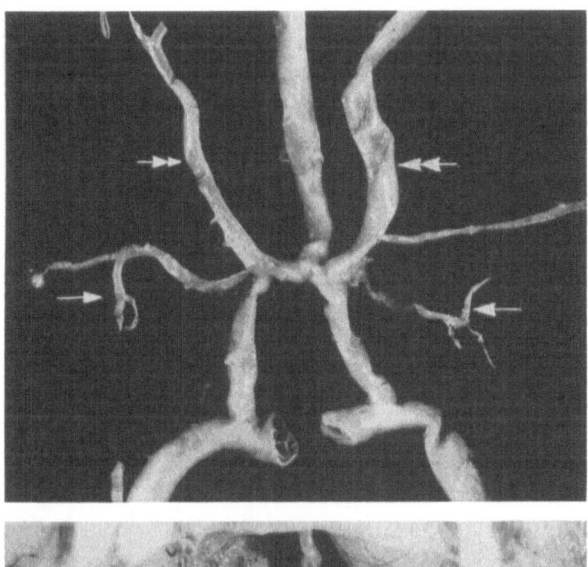

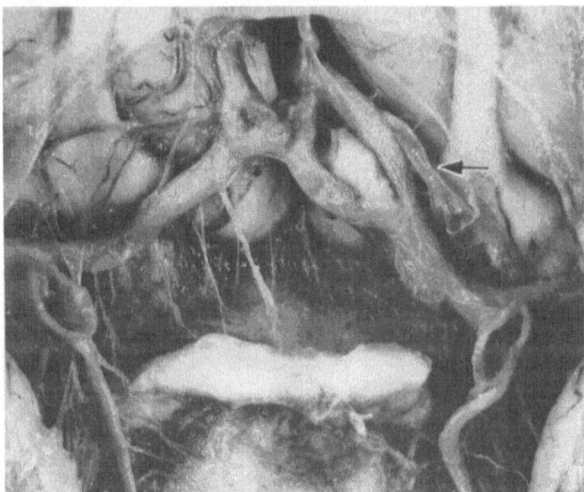

Fig. 2. Variations at the origin of the recurrent artery of HEUBNER. Upper: The recurrent artery (arrows) originates bilaterally at the level of the anterior communicating artery. Between both anterior cerebral arteries (double arrows) there exists an additional A. mediana corporis callosi. Lower: Duplication of the anterior communicating artery with formation of a bridge. In this case the left recurrent artery of HEUBNER (arrow) is unusually well developed

IV. Intracerebral Course and Irrigation Area

Materials and Methods: The supply area of the r.a.H. was studied by means of the injection technique described by SPANNER using an indian ink-Ringer-solution. Injections were performed in a total of 39 cerebral hemispheres. No brains were used in which there was evidence of neurological disease. The average age of patients was 65 years. The dye solution was injected through a polyethylene tube which was directly clamped into the r.a.H. Injection was performed either manually or by the use of an injection apparatus similar to the device described by BEEVOR (1909). By this apparatus, it was possible to vary the injection pressure according to the volume injected per time and, on the other hand, to maintain a constant pressure to avoid vessel rupture and extravasation. In three brains a double injection technique was used. In these cases the supply area of the r.a.H. was stained by injection of cinnabar red-gelatine, while the middle cerebral artery was injected with indian ink. Brains were fixed in 4% formalin and later cut in slices of 5 mm thickness. After photographic documentation, the intracerebral course was further studied by translucent illumination according to the technique of SPALTEHOLZ.

The penetrating branches of the r.a.H. enter the brain by the substantia perforata anterior at the level of the lateral occipital segment of the anterior commissure (Fig. 3). The cerebral cortex around the substantia perforata anterior does not belong to the

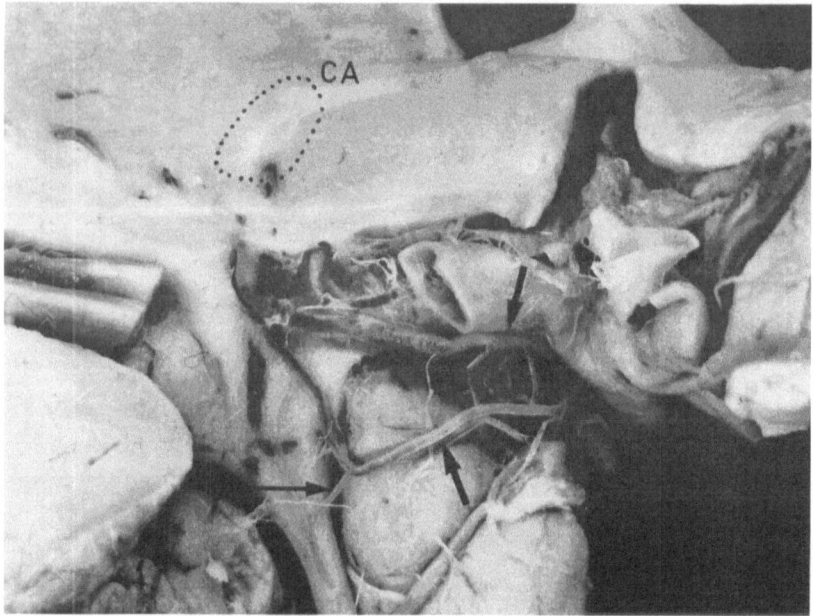

Fig. 3. The recurrent artery of HEUBNER penetrates the brain at the level of the anterior commissure (CA). The small area below the lateral segment of the anterior commissure variably—irrigated by the recurrent artery of HEUBNER is indicated. The recurrent artery of HEUBNER (arrows) is duplicated in this case with the more proximally located vessel providing an extracerebral branch to the tractus olfactorius (small arrow)

irrigation area of the r.a.H. After injection of a dye solution staining was normally restricted to a small area below the lateral part of the anterior commissure (Fig. 3). The most laterally localized segment of the anterior commissure was occasionally stained. There was no staining above and posterior to the anterior commissure. Shortly after entering the brain, the branches of the r.a.H. run in a frontal direction towards the corpus striatum. By translucent illumination it could be demonstrated that most branches first enter the putamen and then cross the internal capsule towards the caudate nucleus. The branches reaching the caudatum represent the intracerebral terminal endings of the artery.

The area within the corpus striatum belonging to the irrigation system of the r.a.H. shows a distinct variability in the individual case. The results are summarized in Fig. 4. There is a *constant supply area* mainly involving the medio-basal third of the corpus striatum. However, in individual cases a far greater extension of the supply area of the r.a.H. could be observed. These extensions beyond the constant supply area may be described as *facultative supply areas* and are demonstrated separately in Fig. 4. In some cases there was also a staining of the fronto-basal cortex and of the subcortical white matter of the frontal lobe. It was possible to demonstrate that this irregular and less intensive staining was caused by extracerebral branches of the r.a.H. Those branches entering the substantia perforata anterior exclusively supplied parts of the rostral basal ganglia. They could never be demonstrated to extend to adjacent areas of the white matter or to the cerebral cortex of the frontal lobe.

V. Infarctions in the Territory of the Recurrent Artery of Heubner

Within an autopsy series of more than 400 cases of cerebral infarctions, only 6 cases of isolated necroses in the territory of the recurrent artery of HEUBNER were found. The main clinical and pathological findings of these cases are summarized in Table 1.

Pathogenesis of Infarctions: In three of the six cases the pathogenesis could be clearly detected. In one of these (case 1) the recurrent artery of HEUBNER was occluded by an atheromatous plaque at the site of its origin from the anterior cerebral artery (Fig. 6). Two other cases were due to rupture of an aneurysm: of the anterior cerebral artery (case 2, Fig. 7 upper) in one and the anterior communicating artery (case 3) in the other. In three cases the pathogenesis of infarction could not be clearly established. One of these (case 4, Fig. 7 lower) revealed bilateral hemorrhagic infarctions localized symmetrically in the medio-basal part of the rostral corpus striatum. This patient suffered from severe circulatory failure; the cerebral arteries, however, did not demonstrate any pathological changes. In case 5 a subarachnoidal bleeding was diagnosed clinically and at autopsy a circumscribed hemorrhage was detected at the base of the corpus striatum. Above this hemorrhage a small but clearly defined infarction was present. The recurrent artery of HEUBNER in its extracerebral course did not demonstrate any occlusive disease but an intracerebral thrombosis could not be excluded. In case 6 there was a patchy arteriosclerosis of the anterior cerebral artery involving the origin of the penetrating arteries but no occlusion of the recurrent artery of HEUBNER could be demonstrated.

Extension of Infarctions: Generally, the infarctions observed were localized within that area of the rostral basal ganglia which was demonstrated by injection specimen to be irrigated by the recurrent artery of HEUBNER. In most cases the infarctions only extended to the area which was shown to be the constant supply area of the recurrent artery of HEUBNER (compare Figs. 4 and 6). In one case the infarction was more extensive in the direction of the supply area of the lenticulostriate branches of the middle cerebral artery.

Clinical Observations

These are summarized in Table 1. The patient reported as case 1 (Fig. 6) died from diabetic coma soon after being admitted to the hospital. No thorough neurological examination was carried out but relatives stated that no neurological disorders had been observed. In cases 2 and 3 the symptoms from the ruptured aneurysm possibly masked minor neurological findings. In case 4 the spastic paresis on the left side was related to an old gun shot wound in the right area centralis.

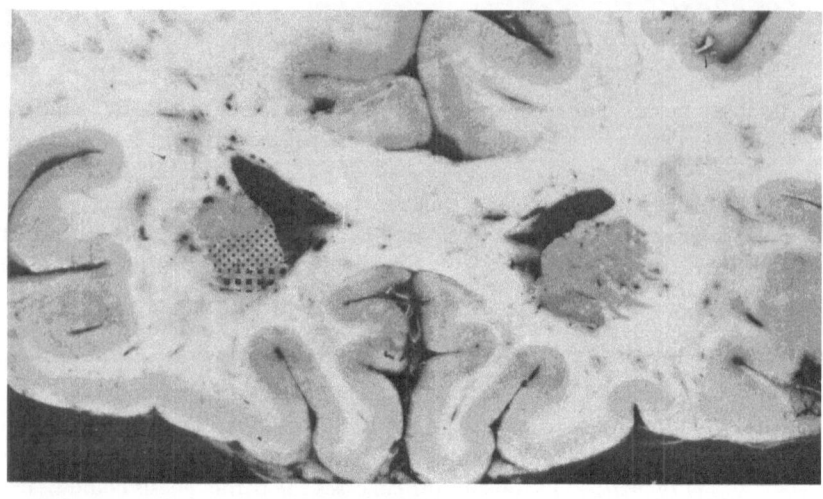

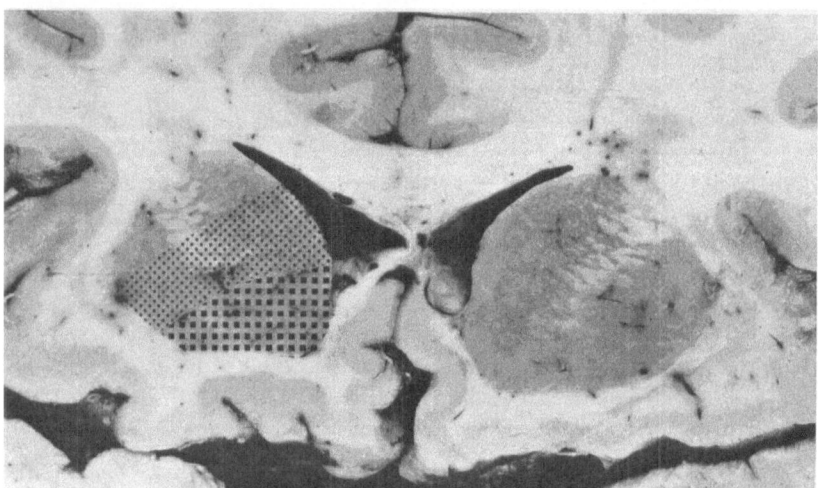

Fig. 4. The irrigation area of the recurrent artery of Heubner as revealed by dye injection. The medio-basal corpus striatum has been demonstrated to represent the constant supply area of the recurrent artery of Heubner (heavy mesh) whereas the adjacent segment above (fine mesh) includes common individual variations (facultative supply area)

Discussion

1. The Anatomy of the Recurrent Artery of Heubner and its Variations

Most authors describing the anatomical features of the recurrent artery of Heubner mention the great variation in the origin of this vessel (Abbie; Almeida; Ayer and Aitken; Critchley).

It was observed that the recurrent artery of Heubner did not arise from the anterior cerebral artery. Other possible sites of origin have been named: the middle cerebral artery (Aitken; Critchley; Almeida), the carotid artery (Critchley) and the anterior choroidal artery (Almeida). However, no details are given regarding the frequencies of these variations. Among 177 brains of the present series the recurrent artery of HEUBNER was proven to originate bilaterally from the anterior cerebral artery in 95%. In one case the intracerebral supply area of the recurrent artery of HEUBNER was irrigated by a perforating branch of the middle cerebral artery. In the remaining cases no definite conclusions could be made. Careful dissection of the penetrating arteries entering the substantia perforata anterior, however, strongly suggested that in these cases the recurrent artery of HEUBNER was also substituted by the most proximal lenticulostriate branch of the middle cerebral artery. Previous data regarding the origin of the recurrent artery of HEUBNER with the anterior communicating artery (a.c.a.) as a reference point are limited. According to WESTBERG, KAPLAN et al. and OSTROWSKI et al., the recurrent artery of HEUBNER originates from the anterior cerebral artery either at the level of the a.c.a. or more proximally. An origin distal to the a.c.a. was found in only 2 of 28 cases by OSTROWSKI et al. SHELLSHEAR (1920/21) describes a distal origin in only one his cases. Our findings demonstrate that if the recurrent artery of HEUBNER did not arise from the anterior cerebral artery at the level of the a.c.a. a distal origin was much more frequent (in 51% of cases on one or both sides) than a proximal one (11% of cases). It is widely accepted that the most characteristic feature of the extracerebral segment of the recurrent artery of HEUBNER is the retrograde course along the anterior cerebral artery (HEUBNER; AYER and AITKEN; KAPLAN and FORD; KOLLMANNSBERGER; WESTBERG). Aside from this, the anatomical variations have been demonstrated to be so numerous and complex that a detailed classification of the extracerebral course — as proposed by KOLLMANNSBERGER — seems to be of restricted use.

Some authors (AITKEN; CRITCHLEY; KAPLAN) have reported that extracerebral branches of the recurrent artery of HEUBNER partially irrigate the tractus olfactorius and the base of the frontal lobe. According to our observations they represent a normal feature rather than an anatomical variation since they could be demonstrated (Fig. 3) with rather regular consistency.

2. Phylogenetic Aspects

During phylogenesis the arteries together with the brain constitute a functional unit (SHELLSHEAR, 1920 to 1921, 1927). As the arteries are constant in their terminal supply, any change in brain structure is reflected in an equivalent alteration in its blood supply (ABBIE). From a phylogenetic point of view ABBIE has interpreted the r.a.H. as "the survivor of the anastomoses over and around the palaeo-olfactorium" between the anterior cerebral artery and the middle cerebral artery (Fig. 5).

Thus, the variations of the origin and the extracerebral course depend on which of these primitive channels remain persistent in the individual case.

Our findings indicating that the supply area of the r.a.H. is most likely irrigated by penetrating branches of the middle cerebral artery, in cases in which there is no origin of r.a.H. from the anterior cerebral artery, are supported by phylogenesis. Extracerebral anastomoses between penetrating branches of both the r.a.H. and the middle cerebral artery would also correspond to this scheme and have been mentioned by KAPLAN but could not be demonstrated in our cases.

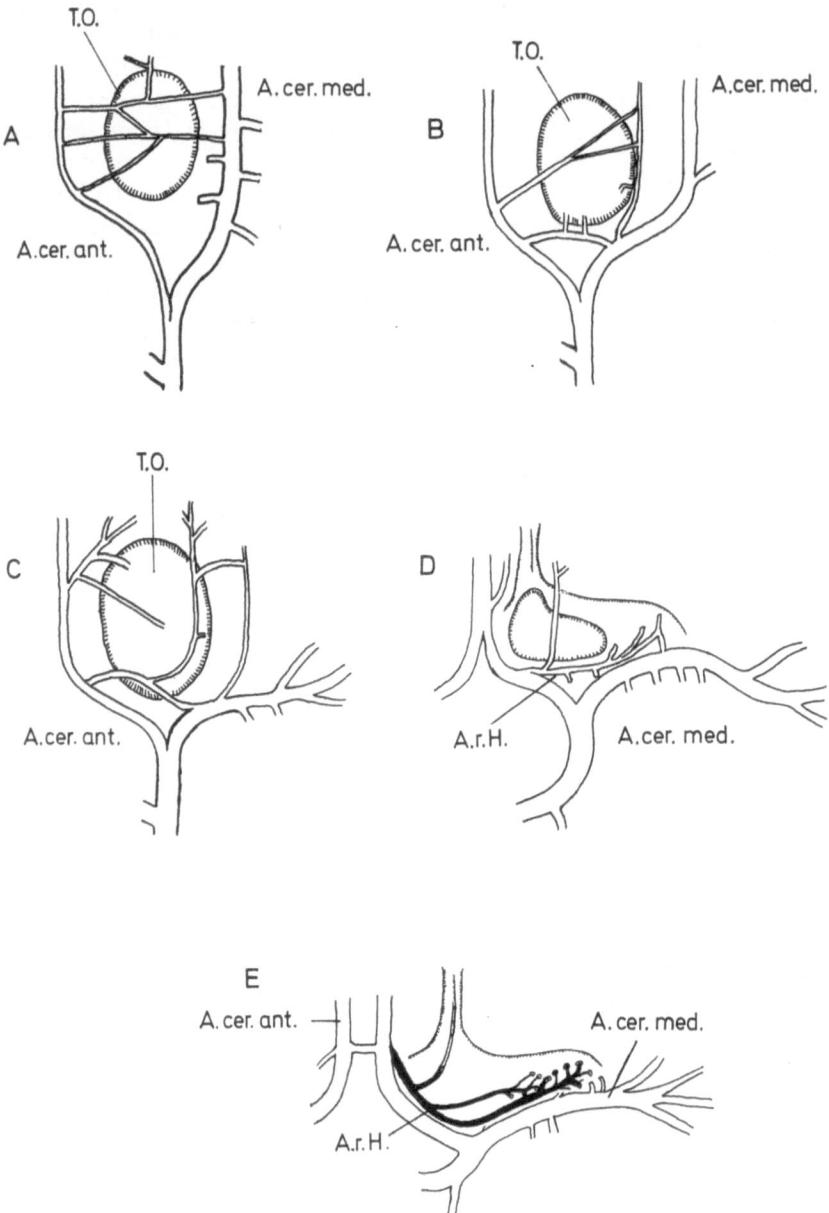

Fig. 5 A—E. On a phylogenetic scale the recurrent artery of HEUBNER can be interpreted as being the remainder of a primitive anastomotic system located at the palaeo-olfactorium and connecting both the anterior and middle cerebral artery. Based on a scheme of ABBIE (1934). A sphenodom, B marsupial (M. thetidis), C ungulate (S. strepsiceros), D primate (C. sabaeus), E homo sapiens, *T.O.* tuberculum olfactorium

Table 1. *Clinical and pathological data of patients with isolated infarctions to the territory of the recurrent artery of* Heubner. *For details on extension and pathogenesis of infarctions, see text*

Case, autopsy No.	Name, age, sex	r.a.H. infarction side	stage[a]	Additional findings of brain autopsy	Clinical observations	Extracerebral disease
1 E 3943	K. Ch., 80, ♀	left	III	Severe arterioscler. of cerebral arteries, cerebral cortical atrophy	No neurological manifestations	Diabetes mellitus, hypertrophy of heart, arteriosclerosis of coronary arteries
2 E 3150	K. M., 57, ♂	right	II	Mass hemorrhage from ruptured aneurysm of the ant. cer. artery, hydroceph. int. et ext. exvacuo	Symptoms of subarachnoidal bleeding	Pneumonia
3 E 3900	K. K., 59, ♂	left	I	Extensive hemorrhage into frontal lobes, ruptured aneurysm of ant. comm. artery	Unconsciousness, symptoms of subarachnoidal bleeding	Hypertensive disease, bronchitis
4 E 3375	R. J., 69, ♂	bilateral	I	Old gun shot wound right area centralis	Left spastic paresis; pain and dysesthesia in both upper limbs, psychomot. restlessness	Vascular failure due to chronic emphysematous bronchitis and pneumonia
5 E 5083	Sch. J., 64, ♂	right	II	Small hemorrhage at the base of the right corpus striatum	5 weeks before death subarachnoid. hemorrhage; pyramidal signs on the left, marked psychomot. restlessness, coma	Hypertensive disease, tuberculosis of the lungs, diabetes mellitus, lung infarctions
6 E 5859	T. B., 68, ♂	left	I	Arteriosclerosis of cerebral arteries	Reduced consciousness, no neurological manifestations	Tuberculosis of the lungs, cor pulmonale, liver cirrhosis, severe generalized arteriosclerosis

[a] According to Spatz (I necrosis, II resorption, III final stage, e.g. cystic scar).

3. The Contribution of the Recurrent Artery of Heubner to the Blood Supply of the Basal Ganglia

Heubner in his original description only mentioned that a part of the caput nuclei caudati was irrigated by the recurrent artery. In later studies the supply area of the r.a.H. has been demonstrated to be more extensive.

Today it is accepted that the medio-basal part of the corpus striatum is irrigated by the r.a.H. Quantitatively, this segment of the corpus striatum varies considerably and therefore to us it seemed to be of little use to give a statistical average. The injection of the dye into the r.a.H. in a total of 39 hemispheres clearly indicated that on a frontal section there is a constant supply area covering about one third of the corpus striatum (Fig. 4) and a facultative supply area covering approximately another third of the corpus striatum.

Ayer and Ayer and Aitken also mentioned parts of the cauda nuclei caudati and the capsula externa as belonging to the supply area of the r.a.H. This could not be confirmed by our observations.

Some authors reported that the r.a.H. also irrigated parts of the rostral globus pallidus (Aitken; Abbie; Critchley; Ostrowski et al.). In our injection specimens staining was restricted to a small area below the lateral segment of the anterior commissure (Fig. 3). From there the intracerebral branches followed a frontal direction towards the corpus striatum. Böhne, in about 50% of his cases, found a separate branch from the anterior cerebral artery which extended through the substantia perforata ant. to the genus capsulae internae which he termed R. genus capsulae internae. Also, Webster et al. in their diagrammatic representation of the distribution of the proximal portion of the anterior cerebral artery illustrate a separate branch supplying the genus capsulae internae and adjacent portions of the putamen and globus pallidus.

4. Infarctions in the Territory of the Recurrent Artery of Heubner

Only a few cases have been reported with isolated infarctions in the territory of the r.a.H. Critchley (1934) was the first to correlate an infarction in the medio-basal corpus striatum with a thrombosis of the r.a.H. Schwarz (1961) in his monograph on cerebral apoplexy demonstrates embolic infarctions, the extensions of which correspond approximately to the irrigation area of the r.a.H. Several cases have been published in the neurosurgical literature (Dandy, 1930 and 1946; Falconer, 1951; McCarty and Cooper, 1951; Gurdjian and Webster, 1953; see also Lazorthes et al., 1956). In each of these cases infarction was caused by clipping one or both anterior cerebral arteries during surgical treatment of aneurysms. The distribution of infarctions was similiar to our cases (with the exception of case 2 of Gurdjian and Webster), however, additional softenings were observed in the cortical supply area of the anterior cerebral artery.

In the present series two infarctions were associated with the rupture of intracranial aneurysms: of the anterior cerebral artery (Fig. 7 upper) and the ant. commu-

---

Fig. 6. Cystic infarction in the territory of the left recurrent artery of Heubner caused by an occlusive atheromatous placque (arrow) at its origin from the anterior cerebral artery. Upper and middle: Posterior-anterior view, lower: Anterior-posterior view (case 1)

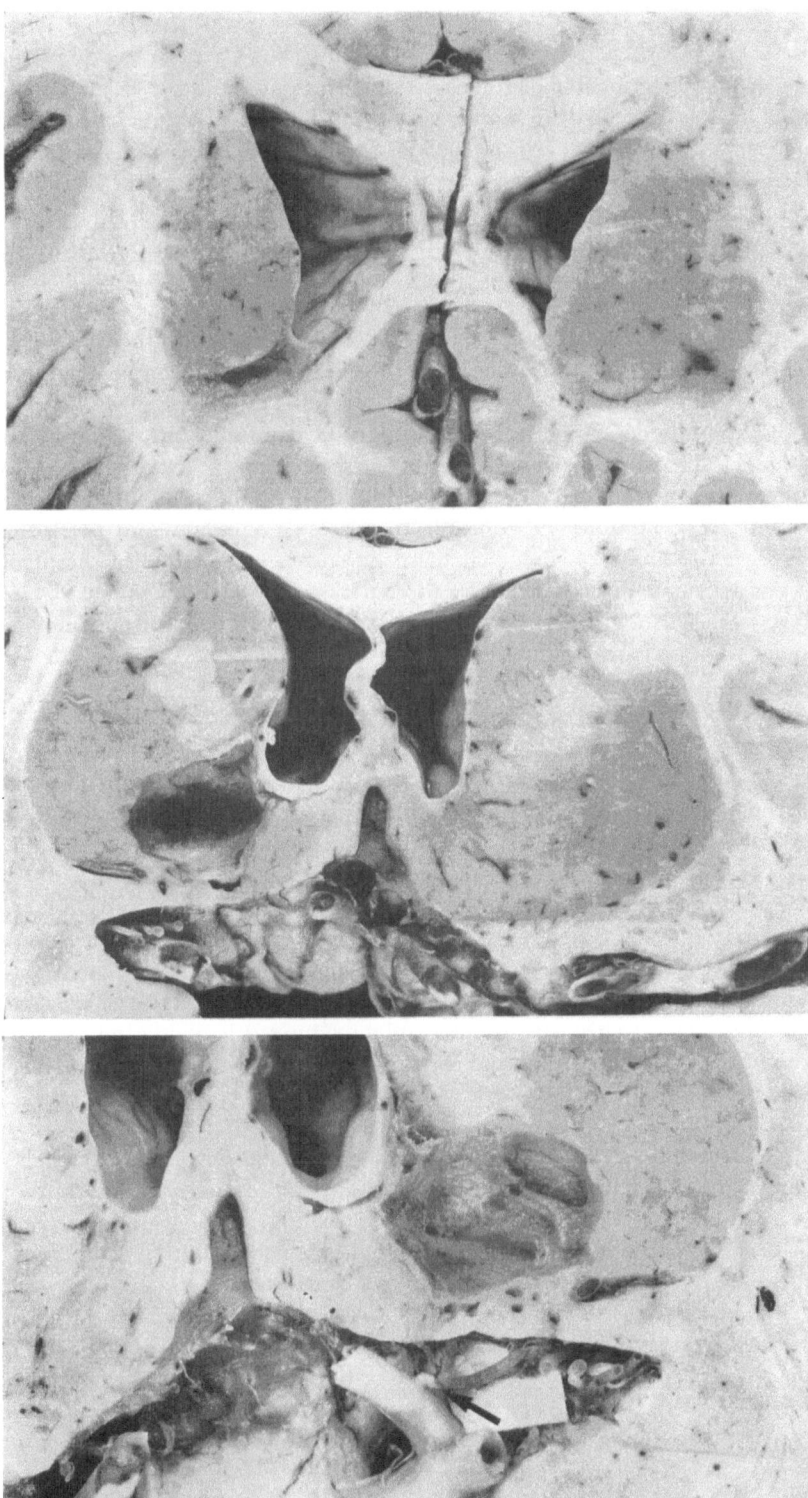

Fig. 6

nicating artery. The possible pathogenesis of such infarctions has been thoroughly discussed by Crompton (1964) and Schneck and Kricheff (1964). Case 1 (Fig. 6) of our series was caused by occlusive arteriosclerosis which, aside from the r.a.H., also involved minor proximal branches of the anterior cerebral artery. In its frontal portion this infarction corresponds well to the supply area of the r.a.H. as revealed by dye injections (Fig. 4) but extends more occipitally including the rostral tip of the globus pallidus (see discussion under 3). Previously, *bilateral infarction* in the territory of the r.a.H. has only been observed after clipping of both anterior cerebral arteries (Dandy; McCarty and Cooper; Gurdjian and Webster). In our case (No. 4, Fig. 7 lower) no local vascular obliteration or stenosis could be detected. The clinical observation of general vascular failure and the bilateral distribution favour the hypothesis that the infarctions might be caused hemodynamically by prolonged fall of blood pressure and low oxygenation of blood (see Zülch; Zülch and Kleihues; Kleihues) although the preference for the territory of the r.a.H. remains unexplained.

From our observations we conclude that isolated infarctions in the irrigation area of the r.a.H. are more frequent than the few cases reported in the literature may suggest. Further, the territory of the r.a.H. is frequently involved in cases of proximal occlusion of the anterior cerebral artery in addition to softenings in the remainder irrigation area of this vessel (Webster *et al.*).

5. Is there a Syndrome of the Recurrent Artery of Heubner?

Dandy (1930, 1946) reported on 10 patients who became unconscious and later died from bilateral ligation of the anterior cerebral artery. He correlated unconsciousness with ischemia and necrosis in the anterior third of the corpus striatum e.g. the territory of the r.a.H. Poppen (1939), on the other hand, demonstrated that the ant. cer. art. could safely be ligated distally to the a.c.a. if the blood pressure was maintained within normal range. In other cases bilateral proximal ligation of the anterior cerebral artery caused marked hyperosmolarity of the blood (McCarty and Cooper; Gurdjian and Webster case 1). The authors suggest that in the cases described by Dandy unconsciousness was also due to electrolyte imbalance (McCarty and Cooper) and that these blood chemistry changes might be a result of ischemic damage to hypothalamic nuclei (Gurdjian and Webster, see also Cooper, 1953) irrigated by proximal penetrating branches of the anterior cerebral artery (Webster *et al.*; Lazorthes *et al.*).

In the case reported by Critchley (thrombotic occlusion of the left r.a.H.) the following symptoms have been related to the infarction found in the ant. third of the corpus striatum: spastic paresis of the right upper limb; slight paresis on the right side of the face, palate and tongue. More often hemiparesis with brachial predominance has been observed in cases of proximal occlusion of the anterior cerebral artery (Baldy; Critchley; Falconer; Webster *et al.*). It should be stressed, however, that clinical symptoms varied considerably and in many cases no neurological disorders were found. If, on the other hand, only the *cortical* supply area of the anterior cerebral artery is involved, a hemiparesis predominantly of the lower limb results from infarction in the paracentral lobule on the medial aspect of the hemisphere.

This is confirmed by the present series (Table 1) which also reveals no constant symptoms. Although there was extensive infarction of the lower portion of the corpus striatum in case 1 (Fig. 6), the patient did not suffer from any noticeable impairment of motor functions. In case 5 there were pyramidal signs but no clear paresis. The patient with bilateral infarction (Case 4, Fig. 7 lower) complained about dysesthesia and pain in the upper limbs but no objective neurological findings could be correlated with ischemia in the territory of the r.a.H.

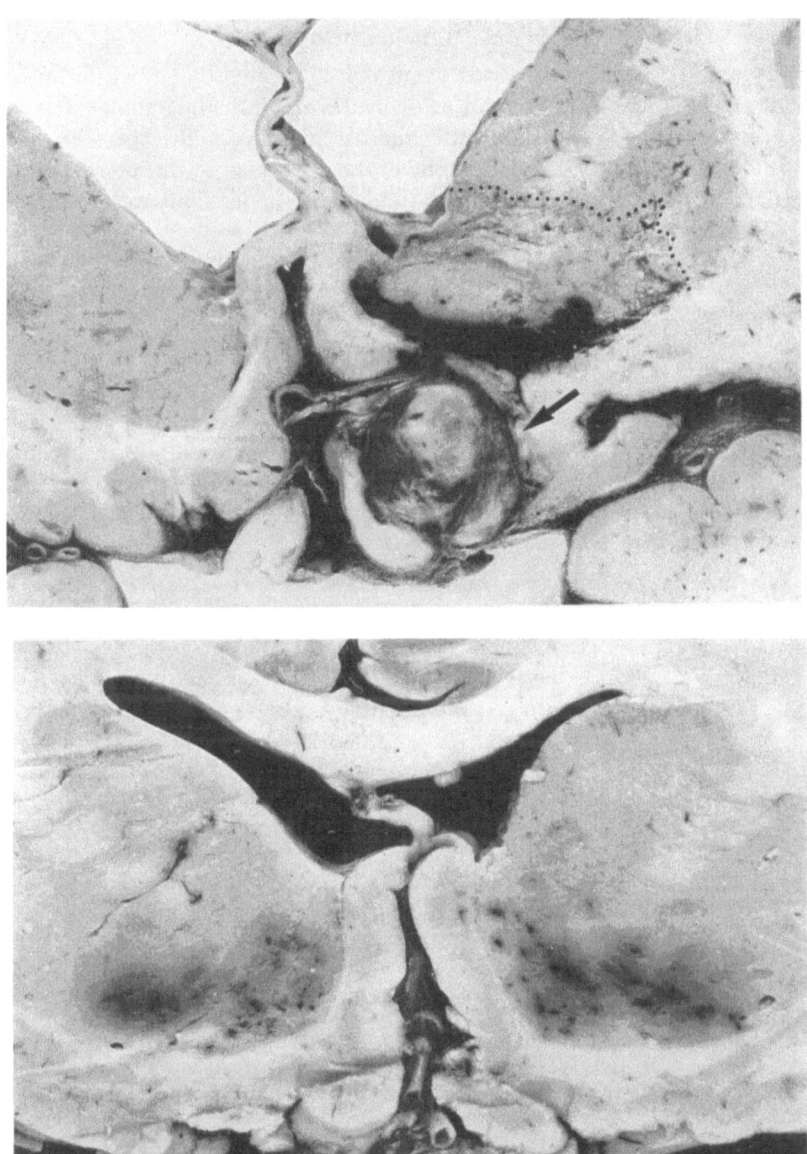

Fig. 7. Upper: Infarction to the irrigation area of the recurrent artery of HEUBNER caused by a ruptured aneurysm (arrow) of the right anterior cerebral artery (case 2). Lower: Bilateral symmetrical infarctions in the territory of the recurrent artery of HEUBNER. No occlusive vascular disease could be detected (case 4)

In 3 of our cases (4, 5 and 6) psychiatric symptoms (psychomotoric restlesness, anxiety) and reduced consciousness were clinically observed. WEBSTER *et al.* described similar findings in patients with proximal occlusion of the ant. cer. artery: "some of the patients with paralysis had associated psychiatric symptoms of insomnia, restlesness, talkativeness and worrisomeness dominating their mood." Although in some of our cases such symptoms were observed, we hesitate to correlate them with circumscribed ischemic lesions in the territory of the r.a.H.

Summarizing the clinical observations it must be stated that a "syndrome of the recurrent artery of HEUBNER" cannot be defined. Only if the infarction extends from the anterior limb to the genus capsulae internae may pyramidal fibers be involved and in this case—based on the topography of the fiber tracts in the internal capsule—contralateral pareses of the face, palate, tongue and upper limb are most likely to occur.

Acknowledgements. We are indebted to Prof. K. J. ZÜLCH for critical suggestions. The technical assistance of Herr H. GÖLDNER is greatly appreciated.

Summary

1. *Origin and extracerebral course of the recurrent artery of* HEUBNER *(r.a.H) were studied in 177 human brains.* In 95% of cases the r.a.H. arose bilaterally from the anterior cerebral artery usually at the level of the anterior communicating artery (in 35% bilaterally). In the remaining cases the origin was more frequently distal, rather then proximal to the anterior communicating artery.

2. *The supply area of the recurrent artery of* HEUBNER *was investigated by means of dye injections in a total of 39 hemispheres.* A constant supply area of the penetrating branches of the r.a.H. could be demonstrated involving the medio-basal third of the corpus striatum. Further, a more extensive facultative supply area is described which includes individual variations of the intracerebral distribution.

3. *The anatomy and pathogenesis of six cases of isolated infarction in the territory of the recurrent artery of* HEUBNER *are reported and correlated with clinical observations.* A syndrome of the r.a.H. cannot be defined. If the infarctions extend posteriorly to the genus capsulae internae then pyramidal fibers may be involved causing a contralateral paresis of brachio-facial predominance.

References

ABBIE, A. A.: The morphology of the fore-brain arteries, with especial reference to the evolution of the basal ganglia. J. Anat. (Lond.) **68**, 433—470 (1934).

AITKEN, H. F.: A report on the circulation of the lobar ganglia. Made to Dr. James B. Ayer. Bost. med. surg. J. **160** (Suppl.) (1909).

ALMEIDA, F. DE: Notes sur l'artère cérébrale antérieure. Bull. Ass. Anat. (Nancy) **28**, 8—10 (1933).

AYER, J. B., AITKEN, H. F.: Note on the arteries of the corpus striatum. Bost. med. surg. J. **156**, 768 —769 (1907).

BALDY, R.: On the distribution of the different arteries supplying the human brain. Phil. Trans. B **200**, 1—55 (1909).

— Les syndromes de l'artère cérébrale antérieure. Thesis (Paris) **1**, 127 (1927).

BÖHNE, C.: Über die arterielle Versorgung des Gehirns; über die arterielle Blutversorgung der subcorticalen Ganglien. Z. Anat. Entwickl.-Gesch. **81**, 151—164 (1926).

COOPER, I. S.: Disorders of electrolyte metabolism in diseases of the central nervous system. Neurology (Minneap.) 3, 119 (1953).

CRITCHLEY, M.: The anterior cerebral artery and its syndromes. Brain 53, 120—165 (1930).

CROMPTON, M. R.: The pathogenesis of cerebral infarction following the rupture of cerebral berry aneurysms. Brain 87, 491—510 (1964).

DANDY, W. E.: Changes in our conceptions of localization of certain functions in the brain. Amer. J. Physiol. 93, 643—658 (1930).

— Location of consciousness center in brain: corpus striatum. Bull. Johns Hopk. Hosp. 79, 34—58 (1946).

DURET, H.: Recherches anatomiques sur la circulation de l'encéphale. Arch. Physiol. (Paris) 6, 60—91, 316—353, 664—693, 919—957 (1874).

FALCONER, M. A.: The surgical treatment of bleeding intracranial aneurysms. J. Neurol. Neurosurg. Psychiat. 14, 155—186 (1951).

GURDIAN, E. S., WEBSTER, J. E.: Blood chemistry studies in bilateral ligation of anterior cerebral arteries. Arch. Neurol. Psychiat. (Chic.) 73, 309—315 (1955).

HEUBNER, O.: Zur Topographie der Ernährungsgebiete der einzelnen Hirnarterien. Zbl. med. Wiss. 10, 817—821 (1872).

— Die luetischen Erkrankungen der Hirnarterien. Leipzig: Verlag F. C. W. Vogel 1874.

KAPLAN, H. A.: Arteries of the brain. An anatomic study. Acta radiol. (Stockh.) 46, 364 to 370 (1956).

— FORD, D. H.: The brain vascular system. Amsterdam-London-New York: Elsevier Publishing-Company 1966.

— RABINER, A. M., BROWDER, J.: Anatomical study of blood vessels in the brain: The perforating arteries of the base of the forebrain. Trans. Amer. neurol. Ass. 79, 38—40 (1954).

KLEIHUES, P.: Über die doppelseitigen, symmetrischen Occipitallappeninfarkte. Dtsch. Z. Nervenheilk. 188, 25—52 (1966).

KOLLMANNSBERGER, A.: Vergleichende Studien über die Arteria Heubneri und die Aa. chorioideae an anatomischen Präparaten, an Korrosionspräparaten und Angiogrammen der Hirngefäße. Morph. Jb. 102, 180—200 (1961).

LAZORTHES, G.: Vascularisation et circulation cérébrales, p. 1—323. Paris: Masson & Cie. 1961.

— GAUBERT, J., POULHES, J.: The central and cortical distribution of the anterior cerebral artery: An anatomical study with neurosurgical implications. Neuro-chirurgie 2, 237 to 253 (1956).

MACCARTY, C. S., COOPER, I. S.: Neurologic and metabolic effects of bilateral ligation of the anterior cerebral arteries in man. Proc. Mayo Clin. 26, 185 (1951).

OSTROWSKI, A. Z., WEBSTER, J. E., GURDJIAN, E. S.: The proximal cerebral artery: An anatomic study. Arch. Neurol. (Chic.) 3, 661—664 (1960).

SCHNECK, S. A., KRICHEFF, I. I.: Intracranial aneurysm: rupture, vasospasm, and infarction. Arch. Neurol. (Chic.) 11, 668—680 (1964).

SCHWARZ, PH.: Cerebral apoplexy—types, causes and pathogenesis. American Lecture Series. Springfield (Ill.): Charles C. Thomas 1961.

SHELLSHEAR, J. I.: The basal arteries of the forebrain and their functional significance. J. Anat. (Lond.) 55, 27—35 (1920—1921).

— A contribution of our knowledge of the arterial supply of the cerebral cortex in man. Brain 50, 236—253 (1927).

SPALTEHOLZ, W.: Über das Durchsichtigmachen von menschlichen und tierischen Präparaten, 2. Aufl. Leipzig: Verlag S. Hirzel 1914.

SPANNER, R.: Die Entwicklung der Darmzotten der Maus durch Knospung und Spaltung untersucht am Gefäßbaum. Morph. Jb. 67, 235—251 (1931).

SPATZ, H.: Pathologische Anatomie der Kreislaufstörungen des Gehirns. Z. ges. Neurol. Psychiat. 167, 301—357 (1939).

TAVERAS, J. M., WOOD, E. H.: In: Diagnostic neuroradiology. Baltimore: The Williams and Wilkins Company 1964.

WEBSTER, J. E., GURDJIAN, E. S., LINDNER, D. W., HARDY, W. G.: Proximal occlusion of the anterior cerebral artery. Arch. Neurol. (Chic.) 2, 19—26 (1960).

Westberg, G.: The recurrent artery of Heubner and the arteries of the cerebral ganglions. Acta radiol. (Stockh.) **1**, 949—954 (1963).

Zülch, K. J.: Pathogenese von Massenblutung und Erweichung unter besonderer Berücksichtigung klinischer Gesichtspunkte. Acta neurochir. (Wien) Suppl. **VII**, 51—117 (1961).

— Kleihues, P.: Neuropathology of cerebral infarction. In: Stroke, Thule International Symposia. (Engel, A., Tage Larsson, Eds.). Stockholm: Nordiska Bokhandelns Förlag 1967.

Clinical Syndromes Due to Ischemia in the Distribution of the Vertebro-Basilar Arterial System*

C. Loeb

Strokes due to ischemia in the vertebro-basilar territory may be classified into three groups: completed strokes due to occlusion of the vertebral and basilar arteries or their branches; completed strokes without demonstrable arterial occlusion; cerebrovascular insufficiency represented by transient or intermittent strokes.

On the basis of the available clinical and angiographic data, such a classification undoubtedly meets with relevant pitfalls.

Firstly, the indication of the angiographic examination has some limitations so that, according to some authors, angiography can be used only in a part of cases (WILLIAMS and WILSON, 1962; SCHOTT et al., 1965; McDOWELL, 1966). Secondly, the interpretation of some arteriographic abnormalities as reliable evidence of occlusion (of posterior inferior cerebellar artery, for instance) is frequently beset with relevant objections (CURRIER, 1962; JANNY et al., 1964; SCHOTT et al., 1965) Thirdly, as I will point out, quite often the topography of the infarction does not correspond to the territory of supply of a given occluded vessel, so that the clinical picture cannot be simply accounted for any given arterial occlusion.

On the basis of these considerations, I thought that it would be better to evaluate the clinical pattern referring—in case of completed stroke—to the suggested anatomical site of the ischemic lesion.

Accordingly, the clinical vascular syndromes of the vertebro-basilar arterial system will be classified under the following two groups: 1. Completed stroke due to (A) isolated or (B) multiple softenings. 2. Cerebrovascular insufficiency: transient and intermittent stroke.

The personal clinical and pathological experience (LOEB, 1962; LOEB and MEYER, 1965) and the most relevant data gathered from the literature will be briefly reviewed.

1. Completed Stroke Due to (A) Isolated or (B) Multiple Softenings

(A) Localized isolated ischemic lesions due to vertebro-basilar disease have their usual location in the medulla oblongata and in the pons.

The well known lateral bulbar syndrome (Wallenberg's syndrome) due to lesions in the lateral bulbar area consists of homolateral facial hyposthesia for pain and temperature, Horner's syndrome, ataxia, paralysis of the soft palate, pharyngeal muscles, vocal chords and, contralaterally, hypoesthesia for pain and temperature over trunk and limbs.

---

* From the Department of Neurology and Psychiatry, University of Sassari, Italy.

Nevertheless, the syndrome shows wide clinical variations ranging from extensive to partial forms.

The lesion may extend rostralward involving the VI and VII cranial nerves, in the caudal direction, involving the XI cranial nerve, and finally in the ventral and medial directions, involving the pyramidal tract with contralateral hemiparesis or hemiplegia.

Partial forms include lesions which occupy only a portion of the lateral area, causing paralysis of the soft palate, pharyngeal muscles, vocal chords, sometimes associated with paralysis of the trapezius and sternocleidomastoideus muscles or facial hemihypoesthesia for pain and temperature.

The well known pontine syndromes were believed by many classical authors to be due to non vascular illnesses. In 1926, Foix and HILLEMAND stated that a crossed pontine syndrome was in itself evidence against a vascular etiology.

Pontine syndromes due to an ischemic lesion may be better represented by a severe hemiparesis or hemiplegia (when the infarction is unilateral) or a quadriparesis or quadriplegia with pseudobulbar and cerebellar signs, when the lesion is bilateral, or a paraparesis or paraplegia in smaller bilateral lesions (see LHERMITTE and TRELLES, 1934; SILVERSTEIN, 1964).

More rarely, multiple and different clinical pictures may be seen, ranging from syndromes of Millard-Gubler to the Foville type when the infarction occurs in the pontine ventral portion at the caudal third or in the pontine tegmentum at the rostral level. Lateral pontine syndromes characterized by disturbances of equilibrium, asynergy, dysmetria, adiadocokinesia homolaterally to the lesion (SILVERSTEIN, 1964) are rather rare and generally associated with paramedian pontine syndromes. Fluctations of the symptomatology deserve particular attention in many cases.

Infarctions in other areas besides the bulbar and pontine ones, just described, are very rare indeed. Isolated midbrain syndromes due to softening such as ventral syndromes (or syndrome of WEBER), dorsal syndromes (or syndromes of the red nucleus) are in fact very rare as pointed out by many authors (HILLER, 1936 and 1952; HASSLER, 1953; BOURDOURES-QUE et al., 1958 etc.). The midbrain softenings described rather recently are really combined midbrain and thalamic softenings (FACON et al., 1958; CASTAIGNE et al., 1962; LHERMITTE et al., 1963).

Isolated ischemic lesions in the cerebellum are also very rare: usually the cerebellar signs are masked by signs due to frequently associated lesions in the brain stem and occipital lobes. Only the case of GOODHART and DAVISON (1936) may be recalled and even in this patient as well the ataxic picture was complicated by other symptoms due to small softenings in the cerebral hemispheres. The symptoms of this case were attributed to an occlusion of the anterior inferior cerebellar arteiy.

Homolateral ataxia was also observed in the unique case of DAVISON et al. (1935) due to superior cerebellar artery occlusion (for more details see in LOEB and MEYER, 1965).

Isolated occipital infarction was demonstrated only in one case of BIEMOND (1951) to be due to basilar occlusion by three silver clips during surgical intervention. Bilateral symmetrical lesions of the occipital lobe are described as an expression of watershed infarction (borderzone between anterior, middle and posterior cerebral arteries) (KLEIHUES, 1966).

Isolated infarctions in the diencephalon and in the temporal area supplied by the vertebrobasilar system were not observed.

Relationships between the topography of the regional softening and the site of occluded vessel. The validity of rejecting a clinical description based on the site of the occluded vessel is proven by the fact that the infarction can be located in an area different from that of the arterial distribution of a given artery. This account should be stressed by calling attention to the well known classification of ZÜLCH (1961) of terminal infarction, central infarction, punctate infarction, watershed infarction (see LOEB and MEYER, 1965).

The lateral bulbar syndrome is due to occlusion of the vertebral artery in about $39^0{}_0$ of cases, to associated occlusion of the vertebral and posterior inferior cerebellar arteries in 26% of cases, to isolated occlusion of the posterior inferior cerebellar artery in 14%, and to isolated occlusion of the bulbar lateral fissure artery in $2^0{}_0$ of cases. No arterial occlusion could be demonstrated in 19% of cases (FISHER et al., 1961). The figures of FISHER et al. (1961) are consistent with the general experience of other authors.

Some other few data may be added. The attempt to make a diagnosis of a specific small arterial occlusion on the basis of the clinical picture is hardly justified considering that, for instance, the posterior inferior cerebellar artery may be absent in $30^0{}_0$ of cases (BAKER, 1961) or may have a very variable area of supply (LOEB and MEYER, 1965). In some cases (see case No. 12 of KUBIK and ADAMS, 1946), a laterobulbar infarction may be due to the occlusion of the anterior inferior cerebellar artery. On the contrary in some cases, the occlusion of one vertebral and of basilar arteries may produce at the medullary level only an ischemic lesion of the restiform body (see case 8 of KUBIK and ADAMS, 1946).

The paramedian pontine softening may be due to occlusion of the basilar artery or to occlusion of the basilar branches (sometimes difficult to demonstrate) or to occlusion of the vertebral artery (see for instance case 68 of YATES and HUTCHINSON, 1961).

In about 40 to $50^0{}_0$ of cases of pontine softenings no vascular occlusion can be demonstrated either at angiography or at autopsy so that the clinical syndrome cannot be ascribed to the occlusion of any given vessel.

(B) More frequently, arterial vertebrobasilar disease gives rise to different and multiple symptoms and signs attributable to ischemic lesion in different areas of the distribution of the arterial system (medulla, pons, midbrain, cerebellum, occipital and temporal lobes, diencephalon) and in various combinations (scattered softenings).

On the basis of the data of literature in 77 verified cases (LOEB and MEYER, in press), scattered softenings due to basilar thrombosis may amount to $65^0{}_0$ of the cases.

The onset and course of the clinical picture may be acute or subacute.

In fact, the clinical onset may be dramatic and fatal within a short period of time (12 h to 2 to 3 days) with a deep coma which hampers a complete neurological examination. The state of coma may progress, recede or wax and wane giving sometimes the possibility to observe some valuable diagnostic signs such as quadriparesis or plegia, dysarthria, visual disturbances and oculomotor paralysis. In these cases, frequently due to embolism (or accidental ligature of the basilar artery during intervention as in the case reported by BIEMOND, 1951), the warning symptoms are very rare.

The great majority of cases show a subacute course. Previous clinical episodes (warning episodes) due to vertebrobasilar insufficiency are present, occurring a few days, months or sometimes years before the final episode, in about 70% of cases (SIEKERT and MILLIKAN, 1955; GAUTHIER, 1963; LOEB and MEYER, 1965), and will be described in detail in the section dealing with cerebrovascular insufficiency.

The symptoms begin suddenly in over half of the cases and are gradually progressive in the remainder. Disorders of consciousness ranging from drowsiness or confusion to true coma are considered as a characteristic feature, being present at the onset in the majority of patients. The state of coma may progress or wax and wane, sometimes exhibiting the feature of the so-called akinetic mutism.

Other common signs are referable to brain stem lesion: hemiplegia or paraplegia, oculomotor paralysis (III, IV, VI cranial nerves), lateral or vertical gaze paralysis, pupillary disturbances. Partial or complete crossed syndromes are seen but are associated with other signs referable to lesions of a different area.

Signs referable to cerebellar lesion are often masked by other signs due to brain stem lesion or cannot be demonstrated because of the comatose state.

Occipital signs are typical including transient cortical blindness, hemianopia, visual agnosia and sometimes anosognosia.

In other words, all signs referable to lesions in different areas supplied by the vertebrobasilar system resulting in various combinations and associations can be observed. It is to be stressed that clear-cut clinical pictures cannot be observed partly because of the almost constant fluctuations of signs both in time and in clinical pattern.

Relationships between the topography of the scattered softenings and the site of occluded vessel. Sometimes the clinical findings seem to be accounted for by a lesion of an area subsequently found to be undamaged at autopsy. The fact is therefore stressed that functional factors as well as anatomical ones must be taken into account in order to explain the clinical picture.

Completed strokes referred to multiple or scattered infarctions are due to basilar thrombosis, or vertebrobasilar thrombosis or bilateral vertebral thrombosis or embolism; in about 40% of cases, however, no arterial occlusion can be demonstrated at angiography or even at autopsy.

Attempts to correlate clinical signs with the area supplied by the occluded vessel are more likely to meet with success when the total or partial occlusion occurs suddenly or when the occlusion is sharply restricted to one segment of the vessel. In this connection attempts to correlate the clinical signs with the site of occluded part of the vertebrobasilar arteries (MARBURG, 1911; DUFFY and JACOBS, 1958; SCHOTT et al., 1965) may be exclusively limited to a few cases in which the infarction corresponds to the anatomical territory of supply of the occluded vessel. Multiple or scattered infarctions occur in various topographical combinations (terminal, watershed and punctate type).

2. Cerebrovascular Insufficiency: Transient and Intermittent Stroke

I consider it advisable to limit the term cerebrovascular insufficiency to the transient or recurrent neurological deficits, implying that the blood supply is transiently unable to meet the metabolic needs, the trouble being not severe or prolonged enough to cause infarction, i.e. a completed stroke.

Cerebrovascular insufficiency includes 1) transient ischemic attacks, i.e. individual or isolated attacks of neurological dysfunction of short duration, sometimes reappearing over a long time course (months or years) 2) intermittent or recurrent attacks or impending stroke, i.e. neurological manifestations occurring at short intervals (hours or days). Progressing stroke or stroke in evolution may be considered as an extension of recurrent episodes since one may develop into the other, as pointed out also by SHAW (1962).

The distinction between the syndrome of cerebrovascular insufficiency and the completed stroke, in view of the fact that the transient or recurring stroke may develop into the completed stroke, is based on the fact that the completed stroke shows no overt evidence of activity or progression.

Transient attacks. They are very rare in the vertebrovascular system compared to the carotid system: 0.8% versus 23.4% based on a total of 589 personal cases of cerebrovascular insufficiency, of which 75 were in the vertebrovascular system.

In our personal cases the signs were: hemianopia; midbrain syndromes; partial lateral bulbar syndromes; cortical blindness.

Table 1. *Symptoms of cerebrovascular insufficiency*

Symptoms	% Frequency
Vertigo	45.5
Visual disturbances	32
Dysarthria	24
Headache	20
Disorders of equilibrium	18
Vomiting	18
Weakness of the legs	12
Disorders of swallowing	9
Drop attacks	5.5
Weakness of one side of the body	3.6
Diffuse dysesthesiae	1.8
Head and eyes deviated	1.8
Auditory changes	1.8

Intermittent or recurrent attacks. The clinical description in this section results from a comparison of 55 episodes in 23 patients personally observed, with the most relevant data of the literature (MILLIKAN and SIEKERT, 1955; WILLIAMS and WILSON, 1962; SCOTT *et al.*, 1965 etc.).

The intermittent attacks show significant differences in symptomatology in the great majority of cases. Only in about 20% of cases can the same clinical features in the different episodes be observed.

The time interval between attacks ranged from a few hours to 1 to 2 weeks.

The episode is characterized by a single symptom in about 20% of cases and by an association of two or more symptoms in the great majority of cases.

The main symptoms are: vertigo, visual disturbances such as diplopia, photopsia, bilateral and transient blindness, dysarthria, suboccipital headache often with vomiting, disturbances in equilibrium, weakness of the legs, dysphagia (see Table 1).

The main signs are: ataxia, paralysis of the VII nerve, visual disturbances (diplopia, hemianopsia), hemiparesis, paralysis of IX and X nerves, dysarthria, and disturbances of consciousness (see Table 2).

Vertigo was the commonest symptom and occurred in 45.5% of cases and was particularly frequent following postural changes; it can sometimes occur as an isolated sign (10%) but more often it is associated with other signs appearing in later episodes in the same case.

Ataxia, sometimes unilateral, is also frequent (40%) but it never occurs alone.

Visual disturbances, including diplopia, perceptual disturbances and blurred vision, appeared in 34% of cases, sometimes (especially hemianopia) as the only sign. Visual perceptual disturbances as vision of bright shapes or wavy lines or black or colored spot were rare.

Pyramidal signs were common (36%): hemiparesis, unilateral hyperactive tendon reflexes; fluctuating hemiparesis appeared in 7%. More rarely the hemiparesis appeared alternatively on each side of the body (3.6%), but only in one case was alternating hemiparesis the only sign.

Table 2. *Signs of cerebrovascular insufficiency*

Signs	% Frequency
Ataxia	40
Paralysis of the VII nerve	40
Hemiparesis	36
Visual disturbances (hemianopia, diplopia, bilateral amaurosis etc.)	34.1
Paralysis of the IX nerve	24
Paralysis of the X nerve	18
Dysarthria	18
Disorders of consciousness	13
Fluctuating hemiparesis	7
Nystagmus	5.5
Alternating hemiparesis	3.6
Hemianesthesia	3.6
Paralysis of the XII nerve	1.8
Paraparesis	1.8
Anesthesia of one side of the face	1.8

Paraparesis was exceedingly rare (1.8%).

Dysarthria and dysphagia occurred in 24% of cases and in one case dysphagia was the only sign.

Headache, which is usually occipital and often associated with nausea and not an isolated symptom, is fairly frequent (20%).

Disturbances of consciousness, ranging from drowsiness to profound coma were present, in our experience, only in 13% of cases.

Akinetic mutism, never observed in our case material, was described by WILLIAMS and WILSON (1962).

Drop attacks were observed in 5.5% and weakness in the lower limbs in 12%.

Symptoms of vertebrobasilar insufficiency may originate from trauma and degenerative changes to the cervical vertebrae, particularly cervical spondylosis and chiropractic maneuvers, due to obstruction of the vertebral arteries (spondylotic vertebral artery compression of SHEEHAN et al., 1960; see also YATES, 1967).

Symptoms of vertebrobasilar insufficiency may be observed in cases of occlusion of the subclavian artery proximal to the origin of the vertebral artery. In such patients the exercise of the upper limb, by increasing blood flow to it, gives rise to a reversed flow in the vertebral artery and blood is "stolen" from the hindbrain to supply collateral circulation to the limb via the distal subclavian (subclavian steal syndrome, REIVICH et al., 1961, or brachial basilar insufficiency NORTH et al., 1962). The symptoms vary widely, ranging from dizziness and disturbance of vision to transient paralysis, aphasia or even olfactory hallucination (YATES, 1967).

Clinical recognition of cases of combined carotid and basilar occlusion or insufficiency is not easy and a high percentage of diagnostic error is to be expected.

In my experience, clinical signs typical of a lesion in the carotid distribution such as contralateral sensory and motor disorders may be associated with signs attributable to a lesion in the vertebro-basilar area such as: severe disturbances of swallowing and nasal voice, with unilateral paralysis of the III, IX, XII cranial nerves.

Despite the frequency of combined caroticobasilar syndromes, clinical diagnosis offers difficulties which are almost impossible to overcome, even though panangiography may provide a valuable help.

As a final comment to this short review, a few physiopathological concepts will be briefly taken into consideration.

The fact that in many cases of total occlusion of the vertebral and/or basilar arteries only partial infarctions of the brain stem and cerebellum are found indicates that an emergency collateral circulation is rapidly available. This collateral circulation may therefore be derived from the carotid system through the posterior communicating arteries and other anastomotic channels already described (LOEB and MEYER, 1965). The factors capable of modifying the cerebral intravascular pressure and the effective collateral cerebral blood flow have been largely discussed by many authors in the past 5 to 10 years (see LOEB and MEYER, 1965). In this connection, the sudden drop in systemic arterial pressure (so called hemodynamic crises) with resulting impairment of the collateral circulation should be considered one of the most relevant factors.

Greater difficulties in physiopathological interpretation arise when arterial occlusion cannot be demonstrated as it happens in approximately 40 to 50% of cases. It is true that the thrombotic occlusion could be dissolved by circulating fibrinolisin, as suggested by serial angiography and studies on blood coagulation (MEYER et al., 1963), but this interpretation cannot meet the whole group of cases, so that several other factors need to be taken into consideration.

Significant stenosis (greater than 50% of the lumen) of extracranial arteries is a fairly frequent finding in non selected series of necroscopies and in many angiographic findings (HUTCHINSON and YATES, 1956; MARTIN et al., 1960; SCHWARTZ and MITCHELL, 1961; McGEE et al., 1962; STEIN et al., 1962; MARSCHALL, 1966) so that its relationship to the occurrence of infarction is not entirely proven in all cases. More recently, however, YATES (1967) points out that stenosis of an artery may reduce the blood flow not only because of the narrowing of the lumen, but also by changing the pattern of flow from a laminar to a turbulent one. Moreover, compression of the vertebral arteries in cases of cervical spondylosis gives rise to permanent or transient neurological defects, supporting the view that narrowing of the lumen of the extracranial arteries has, in some cases, its physiopathological import.

Embolism due to aggregation of platelets, or from release into circulation of fragments of arterial debris originating from atheromatous ulcers or mural thrombi has been taken into account by many authors (McBrien *et al.*, 1963; Loeb and Meyer, 1965; Ross Russell, 1966; Adams, 1966).

There is little doubt that arterial hypertension is probably the primary and precipating factor in some cases of sudden hypertension in chronically hypertensive subjects. Only in this connection may arterial vasospasm possibly be taken into account (Loeb and Roccatagliata, 1964).

The pathogenesis of cerebrovascular insufficiency is similar to that of infarction with the important difference that in insufficiency the reduction in blood supply is less severe and of briefer duration.

A more complete account of the pathogenesis of cerebral infarction and insufficiency may be found elsewhere (Loeb and Meyer, 1965), but I wish here to stress the value of two other factors.

Firstly, the possible occurrence of an intimal swelling of an atherosclerotic artery in analogy to the coronary arteries (Giampalmo and Antoci, 1960). A sudden intimal swelling taking place in the depths of atherosclerotic thickening yields to a reduction of the lumen.

Secondly, stenosis of the small intracerebral arteries may consistently alter cerebral hemodynamics as proved by the experimental research of Meyer and Denny Brown (1957), so that shunting from one territory to another may happen.

The value of an intracranial steal associated with diffuse arterial disease of the small vessels (diabetes mellitus, collagen diseases, hypertensive vasculopathy) must be taken into consideration (as some personal research now in progress seems to prove).

The final conclusion is that infarction and cerebrovascular insufficiency are the terminal event arising from a combination of a series of factors which are both anatomical and functional, and are evidenced differently in each given case.

References

Adams, R. D.: Etude de la nécrose ischémique du cerveau dans l'artériosclérose. Symposium international sur la circulation cérébrale. Paris: Sandoz 1966.

Baker, A. B.: The medullary blood supply and the lateral medullary syndrome. Neurology (Minneap.) **11**, 852—861 (1961).

Biemond, A.: Thrombosis of the basilar artery and the vascularization of the brain stem. Brain **74**, 300—317 (1951).

Boudouresques, J., Roger, J., Bonnal, J., Serratrice, G.: Ramollissements cérébraux. Enc. Méd. Chir., Neurologie, Paris **2**, 17046, B 20 (1958).

Castaigne, P., Buge, A., Escourolle, R., Masson, M.: Ramollissement pédonculaire médian tegmento-thalamique avec ophtalmoplégie et hypersomnie. Rev. neurol. **106**, 357—367 (1962).

Currier, R. D., Schneider, R. C., Preston, R. E.: Angiographic findings in Wallenberg's lateral medullary syndrome. J. Neurosurg. **19**, 1058—1067 (1962).

Davison, C., Goodhart, S. P., Savitsky, N.: The syndrome of the superior cerebellar artery and its branches. Arch. Neurol. Psychiat. (Chic.) **33**, 1143—1174 (1935).

Duffy, P. E., Jacobs, G. B.: Clinical and pathologic findings in vertebral artery thrombosis. Neurology (Minneap.) **8**, 862—869 (1958).

Facon, E., Steriade, M., Wertheim, N.: Hypersomnie prolongée engendré par des lésions bilatérales du système activateur médial: le syndrome thrombotique de la bifurcation du tronc basilaire. Rev. neurol. **98**, 117—133 (1958).

Fisher, C. M., Karnes, W. E., Kubik, C. S.: Lateral medullary infarction. The pattern of vascular occlusion. J. Neuropath. exp. Neurol. **20**, 323—379 (1961).

Gauthier, G.: Contribution à l'étude de la thrombose basilaire. Arch. Neurol. Neurochir. Psychiat. **90/91**, 209—234, 385—411 (1962/1963).

GIAMPALMO, A., ANTOCI, B.: Sugli ematomi e i rigonfiamenti parietali nelle arterie coronarie sclerotiche ristrette quali cause di ulteriore restringimento acuto. Arch. De Vecchi. Anat. Pat. **32**, 485—494 (1960).

GOODHART, S. P., DAVISON, C.: Syndrome of the posterior inferior and anterior cerebellar arteries and their branches. Arch. Neurol. Psychiat. (Chic.) **35**, 501—524 (1936).

HASSLER, R.: Erkrankungen der Oblongata, der Brücke und des Mittelhirns. Erkrankungen des Kleinhirns. In: Handbuch der Inneren Medizin, Neurologie, 5. Bd., 3. Teil. Berlin-Göttingen-Heidelberg: Springer 1953.

HILLER, F.: Die Zirkulationsstörungen des Gehirns und Rückenmarks. In: BUMKE, O., FOERSTER, O., Handbuch der Neurologie, 11. Bd., S. 178—465. Berlin: Springer 1936.

— The vascular syndromes of the basilar and vertebral arteries and their branches. J. nerv. ment. Dis. **116**, 989—1016 (1952).

HUTCHINSON, E. C., YATES, P. O.: The cervical portion of the vertebral artery. A clinico-pathological study. Brain **79**, 319—331 (1956).

JANNY, P., MONTRIEUL, B., PLAGNE, R., CHABANNES, J.: Sur l'étude angiographique de quelques accidents vasculaires du territoire vertébro-basilaire. Rev. neurol. **110**, 58—66 (1964).

KLEIHUES, P.: Über die doppelseitigen symmetrischen Occipitallappen Infarkte. Pathologie und klinisch-ophthalmologische Befunde. Dtsch. Z. Nervenheilk. **188**, 25—52 (1966).

KUBIK, C. S., ADAMS, R. D.: Occlusion of the basilar artery. A clinical and pathological study. Brain **69**, 73—121 (1946)

LHERMITTE, F., GAUTIER, J. C., MARTEAU, R., CHAIN, F.: Troubles de la conscience et mutisme akinétique. Etude anatomo-clinique d'un ramollissement para-médian, bilateral du pédoncule cérébral et du thalamus. Rev. neurol. **109**, 115—132 (1963).

LHERMITTE, J., TRELLES, J. O.: L'artériosclerose du tronc basilaire et ses conséquences anatomocliniques. Jb. Psychiat. Neurol. **51**, 91—107 (1934).

LOEB, C.: Patologia del circolo sottotentoriale. Sist. nerv. **14**, 213—323 (1962).

— MEYER, J. S.: Strokes due to vertebro-basilar disease. Springfield (Ill.): C. Thomas 1965.

— ROCCATAGLIATA, G.: Aspetti clinici dell'insufficienza cerebro-vascolare. Sist. nerv. **16**, 403—415 (1964).

MARBURG, O.: Über die neueren Fortschritte in der topischen Diagnostic der Pons und Oblongata. Dtsch. Z. Nervenheilk. **41**, 41—91 (1911).

MARSHALL, J.: Le role des variations fonctionelles dans les épisodes ischémiques vasculaires cérébraux. Symposium international sur la circulation cérébrale. Paris: Sandoz 1966.

MARTIN, M. J., WHISNANT, J. P., SAYRE, G. P.: Occlusive vascular disease in the extra-cranial cerebral circulation. Arch. Neurol. Psychiat. (Chic.) **3**, 530—538 (1960).

McBRIEN, D. J., BRADLEY, R. D., ASHTON, N.: The nature of retinal emboli in stenosis of the internal carotid artery. Lancet **1963 I**, 697—699.

McDOWELL, F.: Indications for arteriography in cerebrovascular disease. Ass. Res. nerv. Dis. Proc., p. 188—195. Baltimore: Williams and Wilkins 1966.

McGEE, D. A., McPHEDRAN, S. R., HOFFMAN, H. J.: Carotid and vertebral artery disease. A clinicopathologic survey of 70 cases. Neurology (Minneap.) **12**, 848—859 (1962).

MEYER, J. S., DENNY BROWN, D.: The cerebral collateral circulation. 1. Factors influencing collateral blood flow. Neurology (Minneap.) **7**, 447—458 (1957).

— GILROY, J., BARNHART, M. I., JOHNSON, J. F.: Therapeutic thrombolysis in cerebral thrombo-embolism. Double blind evalution of intravenous plasmin therapy. Neurology (Minneap.) **3**, 927—937 (1963).

NORTH, R. R., FIELDS, S. S., DE BAKEY, M. E., CRAWFORD, E. S.: Brachialbasilar insufficiency syndrome. Neurology (Minneap.) **12**, 810—820 (1962).

REIVICH, M., HOLLING, H. E., ROBERTS, B., TOOLE, J. F.: Reversal of blood flow through the vertebral artery and its effect on cerebral circulation. New Engl. J. Med. **265**, 878 (1961).

Ross RUSSEL, R. W.: Les embolies dans l'artériosclerose. Symposium international sur la circulation cérébrale. Paris: Sandoz 1966.

SCHOTT, B., BOURRAT, CH., TRILLET, M., GOUTELLE, A.: Pathologie artérielle du systéme vertébro-basilaire. Rapport de Neurologie; Congrés de Psychiatrie et de Neurologie de Langue Francaise, LXIII Session, Lausanne 1965. Paris: Masson 1965.

Schwartz, C. J., Mitchell, J. R. A.: Atheroma of the carotid and vertebral arterial systems. Brit. med. J. **2**, 1057—1063 (1961).

Shaw, D. A.: The use of anticoagulants in neurology. In: Modern trends in neurology, 3. (Williams, D., Ed.). London: Butterworths 1962.

Sheehan, S., Bauer, R. D., Meyer, J. S.: Vertebral artery compression in cervical spondylosis. Neurology (Minneap.) **10**, 968—986 (1960).

Siekert, R. G., Millikan, C. H.: Studies in cerebrovascular diseases. II. Some clinical aspects of thrombosis of the basilar artery. Proc. Mayo Clin. **30**, 93—100 (1955).

Silverstein, A.: Acute infarctions of the brain stem in the distribution of the basilar artery. Confin. neurol. (Basel) **24**, 37—61 (1964).

Stein, B. M., McCormick, W. F., Rodriguez, J. N., Taveras, J. M.: Post mortem angiography of cerebral vascular system. Arch. Neurol. Psychiat. (Chic.) **7**, 545—559 (1962).

Williams, D., Wilson, T. G.: The diagnosis of the major and minor syndromes of basilar insufficiency. Brain **85**, 741—774 (1962).

Yates, P. O.: The pathological basis for cerebral ischaemia. In: Modern trends in neurology, 4, pp. 180—192. (Williams, D., Ed.). London: Butterworths 1967.

— Hutchinson, E. C.: Cerebral infarction: The role of stenosis of the extracranial cerebral arteries. London: Her Majesty's Stationary Office 1961.

Zülch, K. J.: Genesi e terapia dell'ictus apoplettico. Rass. mens. med. Ted. **3**, 129—142 (1961).

Vertebro-Basilar Occlusion and its Morphological Sequelae*

H. Metzinger and K. J. Zülch

The clinical syndrome of "vertebro-basilar insufficiency" (Millikan and Siekert) is derived from the description of a large series of morphologically controlled clinical descriptions, which were recently summarized by C. Loeb and J. S. Meyer in a brief monograph. Vertebral or basilar thrombosis or stenosis was usually the cause of this insufficiency (most important papers: see Eisenlohr; Hayem; Leyden; Kubik and Adams; Denny-Brown; Cravioto; Biemond and many others).

However, the typical areas of ischemia in the morphological substratum of this syndrome have not yet been sufficiently defined and described, although one may see already a certain stereotyped pattern in the findings of the literature and our own observations. Here we still lack the classical detailed work as has been done for the cerebral arteries (Foix and Hillemand and many others). Particularly detrimental is the fact that angiographic control or additional information is almost lacking in this region.

This study is intended to illustrate the morphological "spectrum" of vertebro-basilar insufficiency by presenting a short description of the typical infarcts including a brief summary of a characteristic case of each type.

1. Dorsolateral Infarcts of the Medulla

This typical infarct, thought to be the substratum of the "Wallenberg"-syndrome, was very rare in our material, although Fisher, Karnes and Kubik were able to describe 16 cases from the Massachusetts General Hospital series, all in cases of vertebral thrombosis. Also Krayenbühl and Yasargil correctly emphasized the fact that this infarct may be caused by *various* vascular lesions due to the variability of the arterial pattern of this region, because a typical posterior inferior cerebellar artery branching from the vertebral artery is not always seen. In fact, sometimes the posterior cerebellar artery may originate from the basilar artery and it can also arise, according to our own dissections, as *two* separate branches (medullary and cerebellar). In their 12 cases clinically observed, a vertebral *thrombosis* was angiographically verified.

The typical infarct (Fig. 1) may involve the sector of the medulla oblongata defined basally by the upper border of the inferior olive, dorsally by a line somewhere below the fourth ventricles or it may even reach the ventricular floor itself. This variation in size is probably due to the differences in the efficiency of the meningeal anastomoses

* From the Max-Planck-Institut für Hirnforschung. Department of General Neurology, and the Neurological Department of the City Hospital Cologne-Merheim (Director: Prof. Dr. K. J. Zülch).

between the middle and superior cerebellar arteries, as shown in Fig. 11 of Loeb and Meyer.

Here in the bulbar region, the discrepancies in the literature concerning the actual supply territories of the arteries of the posterior fossa are particularly remarkable. Dis-

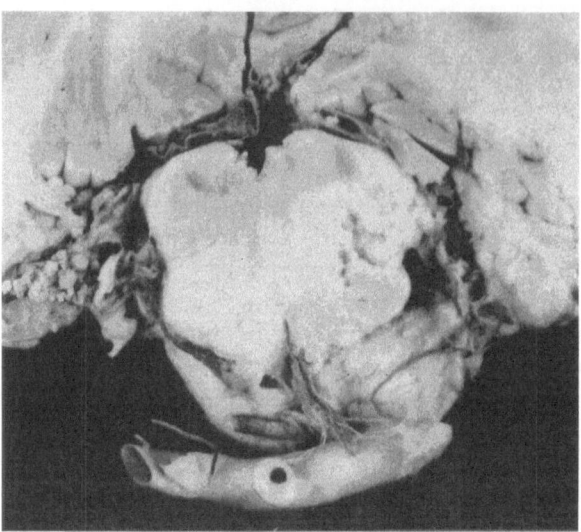

Fig. 1. Typical dorso-lateral infarct of the medulla oblongata (usually the cause of Wallenberg's syndrome)

Table 1. *Topographical distribution of infarcts in 45 cases of vertebro-basilar thrombosis*

 1 case of a dorsolateral medullary infarct
19 cases of paramedian pontine infarcts
 3 cases of short circumferential pontine infarcts
 6 cases of long circumferential pontine infarcts
12 cases of infarcts of the peduncles
       (1 with a rostral extension into the substantia nigra; 2 with a pontine extension)
 9 cases of thalamic infarcts
       (4 thalamo perforating a.; 4 thalamo geniculate a.; 1 atypical location)
15 cases of occipital infarcts
       (4 total, out of which 3 bilaterally symmetrical; 7 proximal/partial; 4 distal/partial infarcts)
16 cases of cerebellar "dorsal" infarcts
12 cases of cerebellar "ventral" infarcts

Anatomical causes:

10 vertebral thromboses
35 basilar thromboses
 2 cases of aneurysm of the basilar and
 1 case of an ectasia ("megadolicho")-basilar artery

regarding this, we only want to point to the dominance of *dorsolateral* infarcts in the medulla oblongata in all series of the literature. Basal necroses, for instance in the pyramids, are a rarity (here we have observed only one little grain like cyst in our great collection of vascular cases). This is due to the supply of this region by the anterior spinal artery.

Our case of a dorsolateral infarct (E 7157) occurred in a male aged 73 with a 90 % stenosis of the ipsilateral vertebral artery. The rest of the lumen showed a fresh thrombosis (Fig. 1).

For the older literature see EISENLOHR; SENATOR; WALLENBERG; FOIX and HILLEMAND; MERRIT and FINNLAND (compiled in HILLER, 1936); moreover M. FISHER, KARNES and KUBIK *et al.*

2. The Paramedian Pontine Infarcts (Paramedian Pontine Arteries)

In our series the most frequent infarct was located in the paramedian segment of the pons (Fig. 2). This infarct was either "total" (8 of 17 cases) or incomplete, either

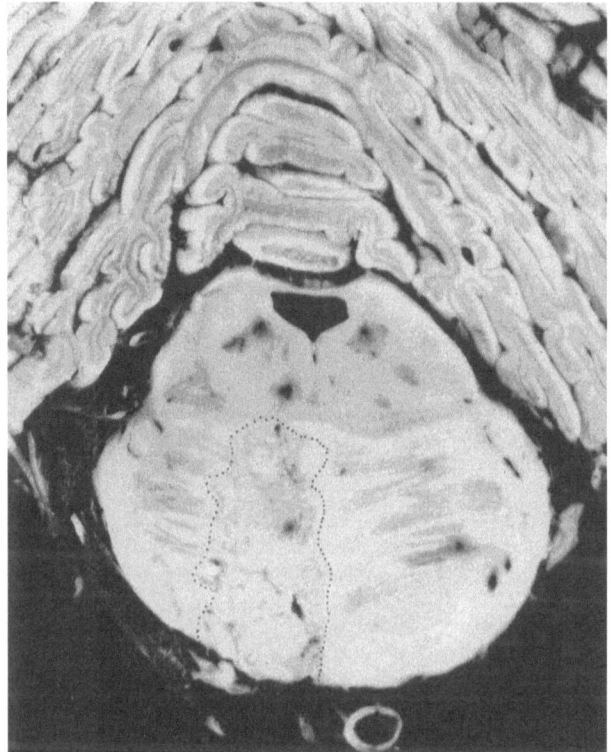

Fig. 2. The most common type of the unilateral "paramedian" infarct of the pons

more basal (2 cases) or dorsal (2 cases), the other parts of the supply territory being spared or the infarct was narrower than the whole zone. Sometimes it was bilateral and rarely combined with other pontine infarcts. 5 of the 17 cases showed an unsystematically distributed necrosis within the supply area.

Usually the infarct was caused by a basilar thrombosis or was due to a local stenosing arteriosclerosis of a paramedian pontine artery. In some cases, the cause could not be clarified.

A summary of the causes of our 17 cases is shown in Table 2.

This is consistent with the findings of LOEB and MEYER and their extensive study of the literature where they discovered 64 cases of basilar thrombosis with a total of 28% paramedian infarcts (for detailed literature see Loeb and J. S. Meyer, their Table 5, p. 86).

An example of a total unilateral infarct of this territory is shown in Fig. 3. It was caused by a high grade stenosis and ectasia of the basilar artery in a lady of 54 years of age.

Table 2. *Seventeen single para-median infarcts caused by*

9 basilar	thromboses
1 vertebral	
7 basilar	stenoses

Apparently there are very different hemodynamic patterns which lead to para-median pontine infarcts. They may depend on the peripontine network of the "menin-geal anastomoses" although there is a complete lack of information about this

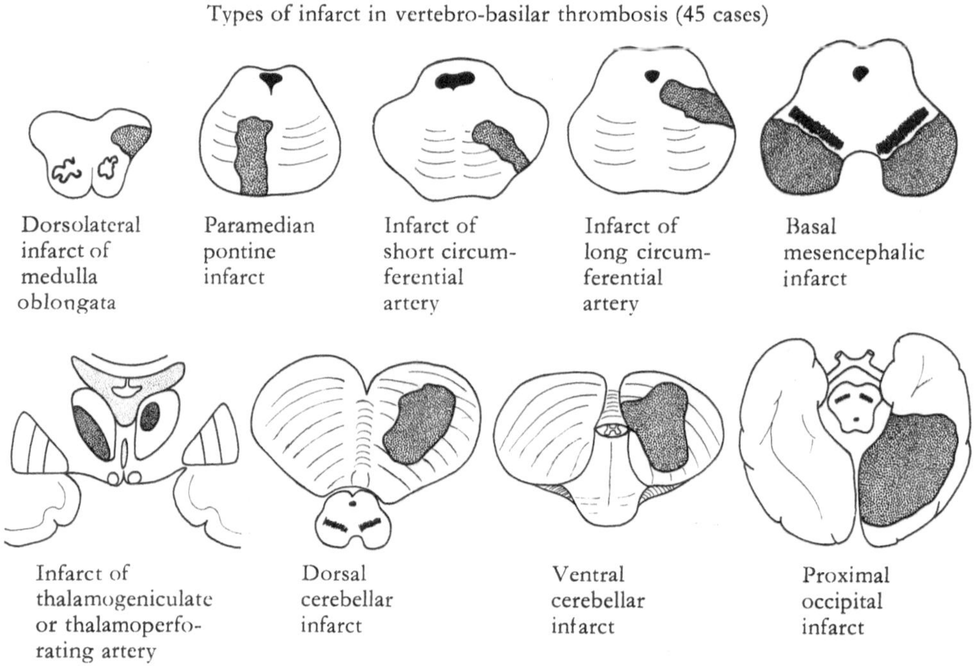

Types of infarct in vertebro-basilar thrombosis (45 cases)

| Dorsolateral infarct of medulla oblongata | Paramedian pontine infarct | Infarct of short circum-ferential artery | Infarct of long circum-ferential artery | Basal mesencephalic infarct |

| Infarct of thalamogeniculate or thalamoperfo-rating artery | Dorsal cerebellar infarct | Ventral cerebellar infarct | Proximal occipital infarct |

Fig. 3. Schematic drawings of the typical infarcts seen in vertebro-basilar insufficiency and/or thrombosis

anastomotic system. Neither morphologically (Van der Eecken) nor neuroradio-logically is there sufficient knowledge about the "collateral" vasculature. Detailed descriptions will be necessary to overcome the discrepancies of the opinions given about the "vascular supply territories" of the pons, before we can understand exactly these pontine infarcts. This will become even more evident in the description of the two locations of the following pontine necroses (short and long circumferential arteries).

Clinically, the paramedian pontine infarct is particularly interesting in the differential diagnosis of the supratentorial (cortical or capsular) hemi-syndromes. Here, for instance, ZÜLCH (1967) pointed to the fact that pontine motor hemisyndromes are often characterised by "proximal" paralyses or pareses, the distal parts being spared or less severely affected (necessary to test isolated movements of the fingers!). This must be due to a dissociation of the pyramidal fibers in the pons, where the bundles for the proximal muscles of the extremities must have a different location.

3. The Latero-basal Pontine Infarcts
(Short Circumferential Arteries — s.c.a.)

More rarely, infarcts in the pons occur in the latero-basal parts supplied by the short circumferential arteries (Fig. 4). These are branches of the basilar artery in the rostral third, according to some authors, while they stem from the superior cerebellar artery according to others. The hemodynamic situation is even less well understood in

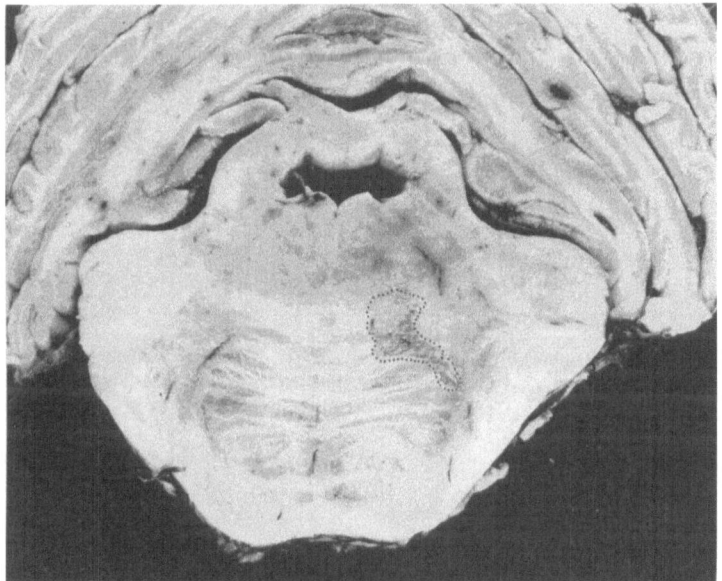

Fig. 4. Small infarct in the supply area of the short circumferential arteries

"ponto-vascular" insufficiency. This is best shown if we try to analyse for instance the extension of the various pontine infarcts in the series of KUBIK and ADAMS. One does not find the clue for the analysis of these patterns even if one considers the action of HEUBNER's meningeal anastomoses and perhaps also SCHMIDT's arachnoidal rings, which are otherwise without great importance in the pathology of infarcts.

The borders of this infarct in the territory of the short circumferential arteries are defined by the lateral three fifths of the pons, a territory which includes some of the pontine nuclei, the lateral borders of the pyramidal pathways and the medial lemniscus, the trigeminal and facial nuclei and the middle cerebellar peduncle.

Three cases of s.c.a. infarcts were combined with paramedian infarcts.

One of our cases (E 6015) was that of a 74 years old lady (Fig. 4) with a rostral thrombosis of the basilar artery. We are not aware yet of a common clinical syndrome which could be correlated with these particular lesions.

4. The Latero-dorsal Pontine Infarcts
(Long Circumferential Arteries = l.c.a.)

These latero-dorsal infarcts (Fig. 5) are located in the pons between the tegmental area and the upper and middle cerebellar peduncles. They are thought to be the sequelae of an occlusion or insufficiency in the long circumferential arteries which are

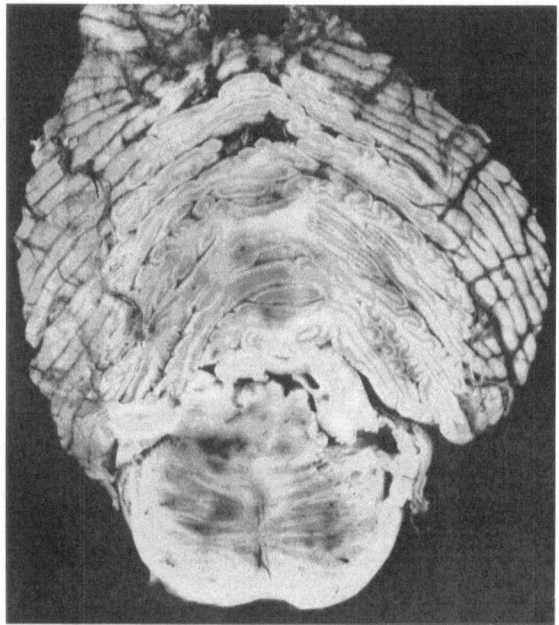

Fig. 5. Large infarct in the territory of a long circumferential artery

assumed to arise from the superior or the middle (inferior anterior) cerebellar artery. Actually these send some branches to the tectal and tegmental regions. The variations in the vascular supply are so remarkable that even good injection specimens have not clarified the question as to where they come from in the majority of cases. Seven of our infarcts in the territory of the l.c.a. were combined with paramedian infarcts.

Our 59 year old patient (E 6607) with a basilar thrombosis had bilateral paramedian and a single latero-dorsal infarcts (Fig. 5) on the side of the major paramedian necrosis, the segment of the short circumferential artery was spared! Also both infarcts were partial and mainly basally situated. This observation again shows the very complex situation of pathological hemodynamics in the pontine region.

Five cases were caused by basilar thrombosis, 1 by an ectatic ("megadolicho"-) basilar artery and one only by a "general" stenosing arteriosclerosis.

5. Mesencephalic (Midbrain-)Infarcts

Typical was the total destruction of the peduncular region (Fig. 6), usually bilateral, the tegmental and tectal segments being spared. Only in 3 cases was there also an extension of the softening into the substantia nigra. Exceptionally rare was the partial infarct, situated in one case only in the substantia nigra without basal peduncular involvement. In two cases only the medial third of the peduncles on one side were necrotic.

In one of the above mentioned larger infarcts a grain sized single cyst in the periaqueductal region was seen.

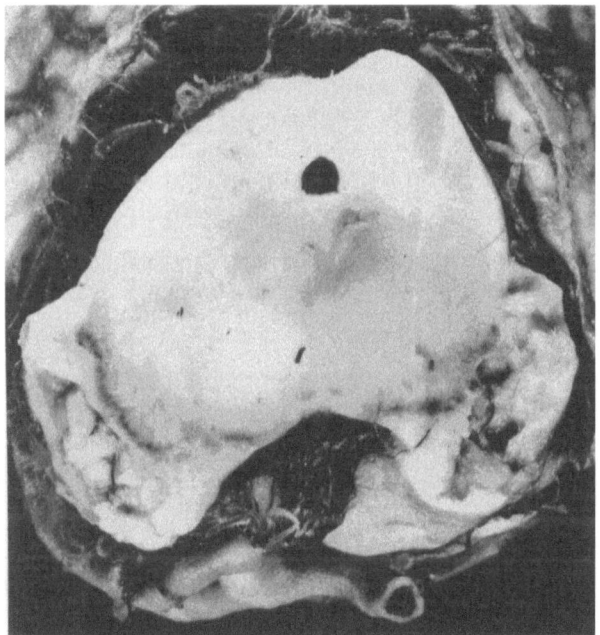

Fig. 6. Bilateral necrosis of the peduncular area (infarct of the pedunculi)

These cases are summarized in the following table.

All mesencephalic infarcts were caused by basilar thromboses.

The zone of destruction in our cases was more frequently bordered dorsally by an intact substantia nigra. This may be explained by the differences in the vascular

Table 3. *Extension of mesencephalic infarcts*

12 mesencephalic infarcts:
    5 cases of bilateral *total* necroses of the peduncles including
      1 with hemilateral extension into the nigra
    4 cases of *unilateral* necrosis also including
      1 with partial softening of the nigra
    2 cases of unilateral *partial* necrosis in the medial third
    1 case of single partial infarct into the nigra

supply, the ventral mesencephalon being supplied by short ("paramedian") branches from the basilar and the proximal part of the posterior artery while the dorsolateral and tectal regions are probably supplied from the superior cerebellar artery. This latter vessel may still have a collateral supply by reflux via the meningeal anastomoses. However, the relative integrity of the dorsal parts in our material remains surprising particularly when compared with the cases of Kubik and Adams where most destructions extended into the tegmental or tectal parts; if one considers the volume of the destroyed and preserved segments of the mesencephalon, an enormous collateral supply must be presumed.

On the other hand, the cases with necrosis only of the medial third of the peduncles are easily explained because the medial supplying vessels arise from the basilar artery, whereas the lateral perforating rami come from the posterior artery. When, as is often the case, the basilar thrombosis extends into the posterior artery only for one or two centimeters, the blood may come into the distal part as well from the "anterior" supply, i.e. the anterior and middle cerebral arteries via Heubner's meningeal anastomoses as also arise *directly* from the posterior communicating artery. Hence the lateral parts will be preserved, whereas the medial supply is blocked by the thrombosis of the basilar artery.

One of our cases is illustrated by Fig. 6 and shows a typical bilateral destruction of the peduncular region combined with a small infarct in the periaqueductal grey matter (male of 56 years, E 6817).

6. Thalamic Infarcts

If the thalamic region is involved by the vascular insufficiency, no *one* typical "thalamic infarct" will be obvious. Our 9 cases were then of different locations. In 4 cases, the "thalamoperforate" arteries were in the location of the infarct, in 4 cases the thalamogeniculate artery. One infarct was atypically situated. In none of our cases did we see infarcts in another supply area as, for instance, in that of the posterior chorioidal artery. Most necroses were smaller than the vascular supply areas described in the text books or publications. A similar problem arises as with the pontine infarcts: why in a particular case does only one artery of the complex supply system undergo insufficiency with subsequent necrosis and why in this particular region? This can hardly be accepted as the action of meningeal anastomoses, at least if we follow the traditional information about the origin of these vessels. This is particularly true if we deny—as we have done up to now—any great influence of the intracerebral *capillary* anastomotic bed in this type of vascular disturbance of the brain, as we had the right to assume (see Zülch, 1969).

Fig. 7 is an example of a thalamic infarct from our series (E 2043 in a 66 year old male). The patient had a thrombosis of the basilar artery probably as a result of a previous arteriosclerotic stenosis. The softening was located in the area of the thalamogeniculate artery and combined with another old infarct in the paramedian segment of the pons.

In summarizing, the thalamic infarcts also belong to the forms of cerebrovascular insufficiency poorly understood, yet may serve as an example of the typical and stereotyped components of the vertebro-basilar insufficiency syndrome.

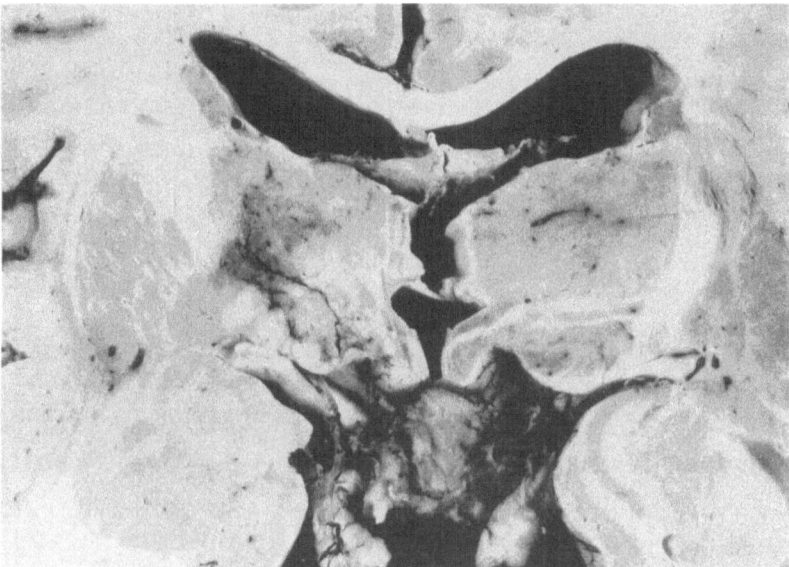

Fig. 7. Infarct in the territory of the thalamoperforate artery

7. Occipital Lobe Infarcts

These infarcts may occur as uni- or bilateral necroses of varying location or extension in the supply area of the posterior cerebral artery. In our laboratory they were subject to a recent investigation by KLEIHUES (1966) and KLEIHUES and HIZAWA (1966), who extended some theoretical considerations to their pathogenesis as published previously by ZÜLCH (1961, 1962).

ZÜLCH described this infarct as an hemorrhagic type caused by a temporal pressure cone (RIESSNER and ZÜLCH, 1939; ZÜLCH, 1961) and later as a special form of an hemodynamic disorder, i.e. the occipital "infarct in the center of a supply area" (which corresponded to the lower parts of the calcarine fissure and leads to a clear cut hemianopia!).

KLEIHUES and HIZAWA were able to prove that several types occurred: the total necrosis of the supply area extending from the hippocampal and thalamic areas up to the occipital pole with sometimes *partial* softenings. This mainly occurred in the *proximal* territory whenever the meningeal anastomoses from the anterior and middle cerebral or even the posterior communicating artery took over the supply of most of the distal borders.

This also took part as the afore mentioned "central" infarct in the territory, when the posterior artery was only stenosed and the bordering parts of the supply territory were supplied by meningeal anastomoses (see Fig. 41, 44, ZÜLCH, 1961). These forms were hemodynamically explained only by an abundant collateral supply from the great meningeal anastomoses of the neighborhood.

This explanation of the pathogenesis fitted well into the clinical findings of either a complete hemianopia (optic radiation) or only the visual hemianopic field defects. In the latter case the central parts were preserved because the pole was spared from infarction due to the action of these collateral anastomoses.

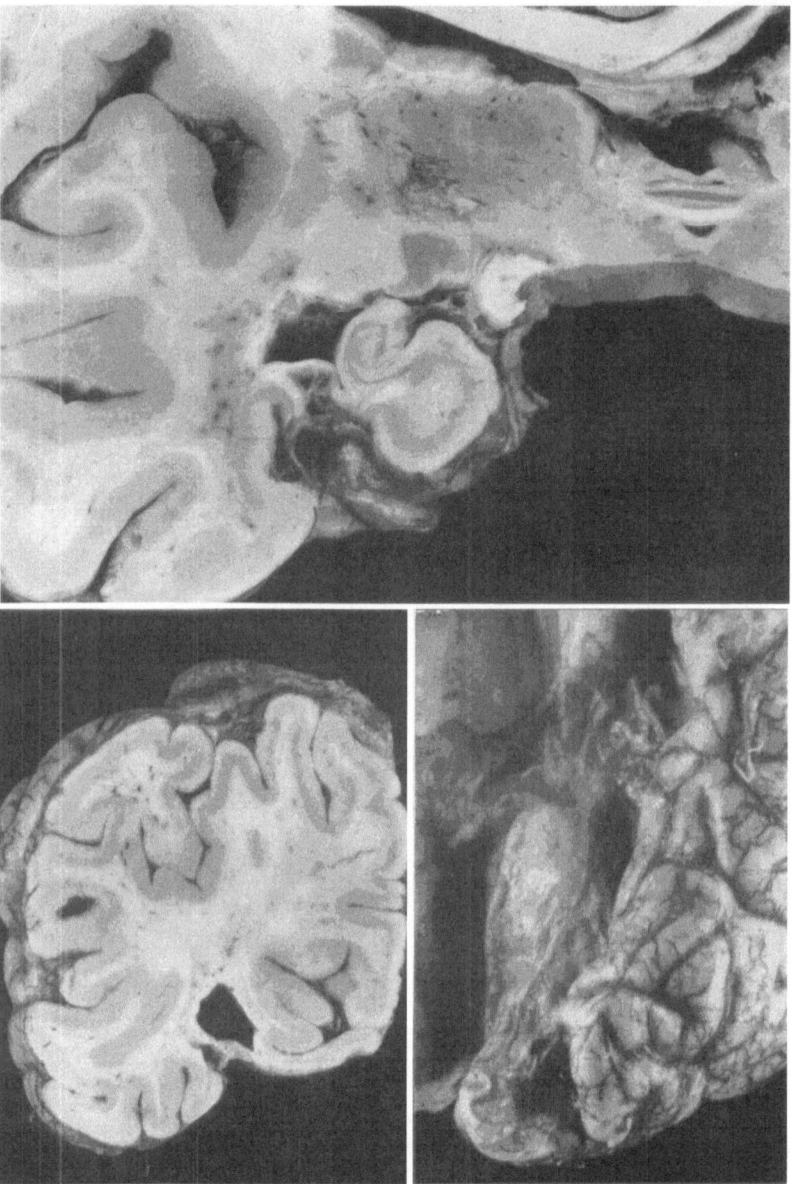

Fig. 8. Infarct of the posterior cerebral artery in a case of proximal occlusion: rostrally, partial necrosis of lateral thalamus, laminar necrosis of Ammon's horn, infarct of the lower "lip" of the calcarine fissure and the occipital pole. (From Kleihues and Hizawa, 1966)

The site and extension of the infarcts in the supply territory of the posterior artery in basilar occlusion is therefore very variable (Fig. 8). Moreover, one has to take into account the fact that every fifth posterior cerebral artery *arises* directly from the carotid artery according to Moniz and many others, i.e. is not part of the vertebro-basilar system at all.

In our total series of 18 we had 3 bilateral symmetrical and 1 unilateral total infarction. The majority of the 14 remaining cases were situated half in the "center of the supply territory" of the posterior cerebral artery and half in a more proximal location.

A typical example (E 5145, Fig. 8) was that of a 60 year old lady with a left sided more proximal infarct of the supply territory of the posterior artery. Included were a partial softening of the pulvinar of the thalamus and a laminar necrosis of the hippocampus. Distally, in the calcarine fissure, only the lower "lip" was severed. As often seen, portions of the occipital pole were spared.

In summarizing: we have to state that this type of infarction is among the most easily understood hemodynamically, when one has a special knowledge of the vascular pattern including the various anastomotic channels and the state of the cardiac and peripheral vasomotor action in a particular case.

8. Cerebellar Infarcts

There were two prominent types of infarcts seen in the cerebellum, i.e. the *"ventral"* and the *"dorsal"* necroses according to the distribution in the main supply territories of the "inferior posterior" (i.p.c.a.) and the "superior" cerebellar artery (s.c.a.).

The typical dorsal infarcts were only rarely situated in the entire supply area in the more distal territory of the i.p.c.a. and usually varied in size. Commonly a typical case shows a *fresh*, mostly hemorrhagic, or *old*, deeply scarred infarct which is located more often in the distal part of the supply area of the s.c.a. (Fig. 9). None of them was "total" or reached far into the depth, i.e. into the white substance including the dentate nucleus.

Out of our series of 16 "dorsal" infarcts, 11 occurred in basilar occlusions, 1 in a basilar aneurysm and 5 in stenosing arteriosclerosis of the basilar artery. In 4 cases they were combined with ventral infarcts while the lateral segments of the cerebellar hemispheres remained free.

The limited extension of these infarcts of the dorsal surface of the cerebellum can only be explained by the apparent rich collateral network of the meningeal anastomoses covering the convexity of the cerebellum, the anastomotic function of which is well proved also neuroradiologically (FIELDS, 1965, 1969, see this book contribution ZÜLCH, Fig. 15, p. 119).

The *"ventral"* infarcts (Fig. 10) on the other hand were usually larger though again by no means "total". The majority of them were old, i.e. scarred and consisted of shallow crater-like grooves located more often in the proximal supply territory and this very likely because of the good anastomotic distal supply. Rarely, when an infarct became "total" in the cerebellar ventral area, i.e. the territory of the i.a.c.a., a corresponding *dorsolateral medullary* infarct was never seen in our cases.

Moreover, there were variations in the site which were hard to understand when most of the usual supply territory was spared and the main extension was "lateral". We were not able to detect whether these and some other of the "atypical" infarcts were located in "frontier zones" or "watershed" areas of the three great cerebellar arteries. Some cases showed multiple necrotic foci in the area of the i.p.c.a. or occurred in combination with "dorsally" located necroses.

In one case there was only an old, ovoid, cystic infarct in the depth lateral to the dentate nucleus; this could be understood as a "terminal" infarct of the s.c.a. (E 1730).

Of the 22 ventral infarcts only 4 were entirely atypical. Pathogenetically, 4 of the cerebellar infarcts occurred in vertebral thrombosis, 5 in basilar thrombosis, 1 in a case of

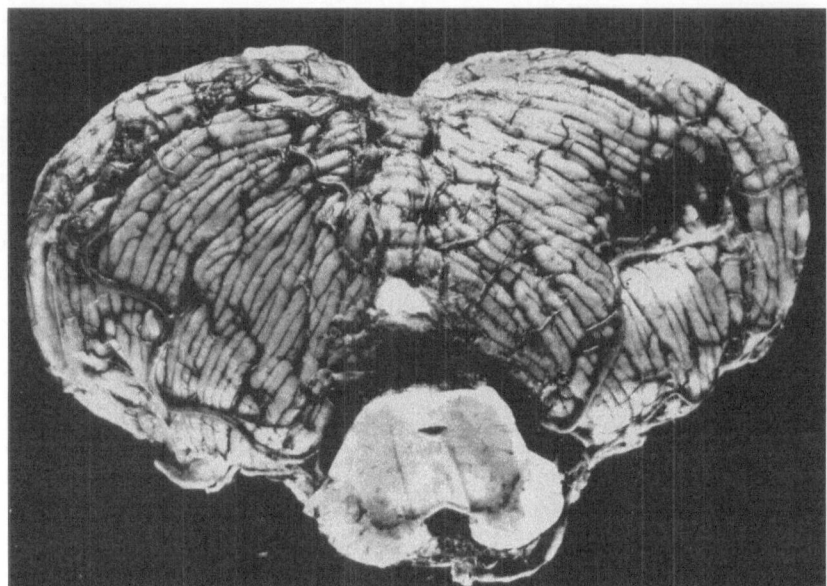

Fig. 9. Small "dorsal" cerebellar infarct

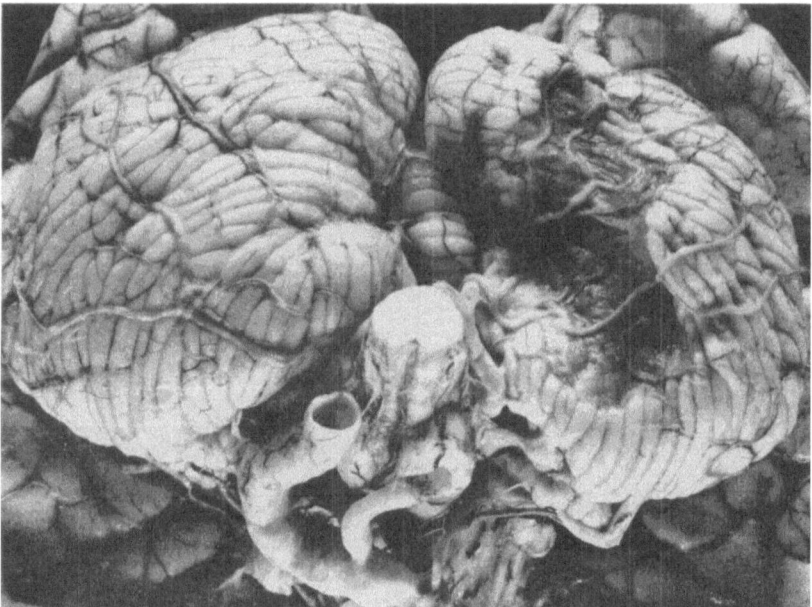

Fig. 10. Median sized "ventral" cerebellar infarct

basilar aneurysm (see Fig. 10) and in the 12 remaining cases only a stenosing type of arterio-sclerosis was present.

A typical case of ours is demonstrated by Fig. 10. The patient was a 71 year old female (E 5486) with a cerebellar infarct caused by vertebral thrombosis and an ectatic type of basilar arteriosclerosis. A close clinical correlation was not possible though there was a history of cerebellar episodes.

In *summarizing* these various infarct patterns of the great morphological syndromes in vertebro-basilar insufficiency, we want to describe *one* of our cases which had a greater number of infarcts concomitant with a fairly full blown syndrome.

This was a 57 year old male who had intermittent visual disturbances in the last two weeks prior to admission, moreover double vision, dysarthric speech disturbances and finally an "epileptic status" with vomiting; bilateral Babinski sign; BP 140/90.

The clinical picture was characterised by unconsciousness, tonic-clonic "fits" of all four extremities of the mesencephalic type of "decerebration". No success was obtained by anticonvulsant medication. CHEYNE-STOKES type of breathing necessitated artificial respiration. Beginning the 8th day, an "apallic syndrome" (akinetic mutism) developed. Death occurred on the 65th day.

Autopsy: 90% stenosis of the arteriosclerotic basilar artery in the middle segment with occlusion by an older thrombus which was consistent with the duration of the clinical syndrome. Extension into the right posterior cerebral artery on the left side only to the point of the posterior communicating artery. A marked stenosing arteriosclerosis of the other basal cerebral arteries was also present. Infarcts: Extensive destruction of the mesencephalon with sparing only of the lateral parts and periaqueductal tectum. The whole pons was included so that the infarct ended 5 mm rostral to the medulla oblongata. Moreover, grainlike cysts were present in the right thalamus and there was an almost total right occipital infarction. Laminar necrosis in the Ammon's horn, a small fresh left occipital infarct, and a small left ventral cerebellar infarct were also noticed.

In summarizing: Basilar thrombosis was associated with extensive infarction of the mesencephalic and pontine basal region, thalamic, cerebellar and occipital infarcts. The clinical syndrome of "decerebration" correlated well with these findings.

Comments

This study shows that the knowledge of the vascular supply of the lower brain stem and its anastomotic channels remains poor and that a complete understanding of the patterns of infarcts and the clinically defined morphological zones of vertebro-basilar vascular insufficiency is not yet possible. Their pathogenesis from the point of hemodynamics remains often questionable, in contrast to the syndromes of the three major *cerebral* arteries.

However, the value of this study seems to lie in the description of "typical" infarcts, i.e. those which occur most frequently in a great series of vertebro-basilar stenosing or occluding arterial disease.

This is true for some types of vertebro-basilar infarcts, which are easy to analyse as, for instance, the occipital infarcts where the modern hemodynamic concept (ZÜLCH; KLEIHUES and HIZAWA) can be successfully applied.

There must be some peculiarities in the flow within the vertebro-basilar system, which are not yet sufficiently understood. The neuroradiologist is familiar with laminar flow in the basilar artery, when the contrast medium is injected into *one* vertebral artery. He also knows that in these cases the *ipsilateral* posterior c.a. is filled with a certain predilection.

Apart from more detailed morphological flow studies within this region, which is hidden angiographically by the bony shadows of the base of the skull, we are in great need for new morphological investigations which should be based on a large number of specimens, i.e. give data about the *relative frequency of the vascular supply by a certain artery*.

Correlative studies about the pathology, pathogenesis and clinical syndromes of vertebro-basilar disease are still of great importance, because the correlation between the neurological syndromes and the pathology is still pretty loose.

Summary

The results of a morphological study of 45 cases of vertebro-basilar thrombosis and their typical infarcts are described. The vascular supply territories are discussed and correlation to the infarct patterns has been attempted. The great variability of the vascular and anastomotic pattern within the vertebro-basilar system is emphasized and some probable peculiarities of flow are discussed.

References

BIEMOND, A.: Thrombosis of the basilar artery and the vascularization of the brain stem. Brain 74, 300—317 (1951).

CRAVIOTO, H.: Occlusion of the basilar artery. A clinical and pathologic study of 14 autopsied cases. Neurology (Minneap.) 8, 145—157 (1958).

DENNY-BROWN, D.: The treatment of recurrent cerebrovascular symptoms and the question of "vasospasm". Med. clin. N. Amer. 35, 1457—1474 (1951).

— FOLEY, J. M.: The syndrome of basilar aneurysm. Trans. Amer. neurol. Ass. 77, 30—34 (1952).

EISENLOHR, L.: Über Bulbär- und Ponsaffektionen. Arch. Psychiat. Nervenkr. 9, 1—48 (1879).

FIELDS, W. S.: Stenosis of internal carotid artery. World Congresses of Neurological Sciences, Joint Sessions, New York Sept. 1969.

— BRUETMAN, M. F., WEIBEL, J.: Collateral circulation of the brain. Monogr. in the surgical sciences, Vol. 2, 183—259. Baltimore: Williams & Wilkins Comp. 1965.

— RATINOV, G., WEIBEL, J., CAMPOS, R. J.: Survival following basilar artery occlusions. Arch. Neurol. (Chic.) 15, 463—471 (1966).

FISHER, C. M., KARNES, W. E., KUBIK, C. S.: Lateral medullary infarction. The pattern of vascular occlusion. J. Neuropath. exp. neurol. 20, 323—379 (1961).

FOIX, CH., HILLEMAND, P.: Rôle de l'oblitération incomplète et du spasme dans la pathogénie de certaines hémiplégies. Soc. Méd. des Hop., 7. Nov. 1924.

— — Les artères de l'axe encéphalique jusqu'à diencéphale inclusivement. Rev. neurol. 32, 705—739 (1925).

HAYEM, G.: Sur la thrombose par arterite du tronc basilaire comme cause de mort rapide. Arch. Physiol. norm. Path. 1, 270—289 (1868).

HEUBNER, O.: Zur Topographie der Ernährungsgebiete der einzelnen Hirnarterien. Zbl. med. Wiss. 10, 817—821 (1872).

— Die luetische Erkrankung der Hirnarterien. Leipzig 1874.

HILLER, FR.: Die Zirkulationsstörungen des Gehirns und Rückenmarks. Handb. Neurol. II/3, S. 178—465. Berlin: Springer 1936.

— The vascular syndromes of the basilar and vertebral arteries and their branches. J. nerv. ment. Dis. 116, 988—1016 (1952).

KLEIHUES, P.: Über die doppelseitigen symmetrischen Occipitallappen-Infarkte. Pathogenese und klinisch-ophthalmologische Befunde. Dtsch. Z. Nervenheilk. 188, 25—52 (1966).

— HIZAWA, K.: Die Infarkte der A. cerebri posterior: Pathogenese und topographische Beziehungen zur Sehrinde. Arch. Psychiat. Nervenkr. 208, 263—284 (1966).

KRAYENBÜHL, H., YASARGIL, M. G.: Die vaskulären Erkrankungen im Gebiet der A. verte-bralis und A. basilaris. Stuttgart: Thieme 1957.

— — Die zerebrale Angiographie. Stuttgart: Thieme 1965.

KUBIK, CH. S., ADAMS, R. D.: Occlusion of the basilar artery. A clinical and pathological study. Brain 69, 73—121 (1946).

LEYDEN, E.: Über die Thrombose der Basilar-Arterie. Z. klin. Med. 5, 165—185 (1882).

LOEB, C., MEYER, J. S.: Strokes due to vertebro-basilar disease. Springfield (Ill.): Charles C. Thomas 1965.

MERRIT, H., FINNLAND, M.: Vascular lesions of the hind brain. Brain 53, 290—311 (1930).

MILLIKAN, C. H., SIEKERT, R. G.: Studies in cerebrovascular disease: I. The syndrome of intermittent insufficiency of the basilar arterial system. Proc. Mayo Clin. 30, 61—68 (1955).

MONIZ, E.: Die cerebrale Arteriographie und Phlebographie. In: Ergänzungsband II zum Handbuch der Neurologie (BUMKE, O., FOERSTER, O., Eds.). Berlin: Springer 1940.

RIESSNER, D., ZÜLCH, K. J.: Über die Formveränderungen des Hirns (Massenverschiebungen, Zisternenverquellungen) bei raumbeengenden Prozessen. Dtsch. Z. Chir. 253, 1—61 (1939).

SCHMIDT, H. W.: Über Embolien in den Arterienkreisen der Pia mater. Z. exp. Med. 125, 401—408 (1955).

— Über Arterienkreise in der Pia mater des Menschen. Dtsch. Z. Nervenheilk. 172, 526 bis 530 (1955).

SENATOR, H.: Apoplektische Bulbärparalyse mit wechselnder Empfindung. Arch. Psychiat. Nervenkr. 11, 713—721 (1881).

— Zur Diagnostik der Herderkrankungen der Brücke und des verlängerten Markes. Arch. Psychiat. Nervenkr. 14, 643 (1883).

VAN DER EECKEN, H.: Discussion of „Collateral circulation of the brain". Neurology (Minneap.) 11, 16—19 (1906)

— De Anastomosen Tussen de Leptomeningiale Slagaders van het Encephalon. Proefschrift Rijksuniversiteit te Gent. Gent: N. V. Drukk. Erasmus Ladeberg.

— ADAMS, R. D.: The anatomy and functional significance of the meningeal arterial anastomoses of the human brain. J. Neuropath. exp. Neurol. 12, 132—157 (1953).

WALLENBERG, A.: Akute Bulbäraffektionen. Arch. Psychiat. Nervenkr. 23, 504—517 (1895).

— Klinische Beiträge zur Diagnostik akuter Herderkrankungen des verlängerten Markes und der Brücke. Dtsch. Z. Nervenheilk. 19, 227—248 (1901).

ZÜLCH, K. J.: Die Pathogenese von Massenblutung und Erweichung unter besonderer Berücksichtigung klinischer Gesichtspunkte. Acta neurochir. (Wien), Suppl. VII, 51 bis 117 (1961).

— Neuere Anschauungen über die Entstehung der cerebralen Insulte. (Extrait du Livre du Dr. LUDO VAN BOGAERT). Acta Medica Belgica, Brüssel, 890—904 (1962).

— Zur Pathogenese des cerebrovasculären Insultes. Internist 4, 64—70 (1963).

— Morphology and Pathogenesis of Cerebral Infarction. Anales del XII Congreso Latinoamericano de Neurocirugia, Symposium International de Investigaciones Neurologicas S. 255—265. Lima, 15—19. 10. 1967.

— Allgemeine Prinzipien bei der Entstehung der Kollateralkreisläufe der Hirnarterien. Radiologe 9, 396—406 (1969).

— Reconsiderations of the clinical problem of cerebrovascular insufficiency. In: Research on the cerebral circulation, S. 1—41. III. Intern. Salzburg Conference. (MEYER, J. S., LECHNER, H., EICHHORN, O., Eds.). Springfield: Charles C. Thomas 1969.

The Vascular Syndrome
of the Parieto-Temporo-Occipital "Triangle"
Based on 18 Cases

H. Gastaut, R. Naquet, and R. A. Vigouroux

The importance of the parieto-temporo-occipital (p.t.o.) "triangle" is well known in vascular pathology. This region corresponds, as far as is known, approximately to the cortico-subcortical projections of the ventricular trigone and is situated at the border-line of three arterial supply territories, i.e. of the middle, anterior and posterior cerebral arteries. The vascular supply of this region, the boundary zone of the three arterial systems, is protected against circulatory insufficiency by an auxiliary supply from neighbouring vessels. Numerous anatomico-physiological investigations, as well as clinical and radiological papers, have emphasized the significance of these auxiliary systems (Adams and Van der Eecken, 1953; Fischer, 1958; Alajouanine et al., 1959; Zülch and Behrend, 1961; Baker et al., 1963) and have particularly stressed the importance of the pial anastomoses which unite at the cortical surface and thus interconnect the terminal branches of the three main arterial systems.

Theoretically, this kind of vascularisation ought to protect this region against vascular accidents. Yet, various authors have stressed the frequency of infarcts in this region, which may be multiple or very small and patchy [Alajouanine et al., 1957, 1959 (1, 2), 1960; Zülch and Behrend, 1961]. Similarly, Naquet (1965) and Naquet et al. (1965) have emphasized the importance of this parieto-temporo-occipital (p.t.o.) triangle for the origin of epileptic activity based on its vascular pathology. Finally, two of us (Gastaut and Naquet, 1965) have, in contrast to the reports of classical vascular brain syndromes originating from one artery, described the diffuse "extraterritorial" (borderline) syndromes of the cortex which have a topographical predilection for the p.t.o. region.

Two hypotheses can explain this topographical predilection: First, because of its *distal* location (with regard to the origin of its supply arteries), the p.t.o. "triangle" shows a functional vulnerability, the more so because the tortuosities in the actual course of the middle cerebral artery actually triple the distance to this site; and moreover, because anastomoses at the angular gyrus are relatively sparse (Alajouanine et al., 1959). A developmental defect producing an insufficiency or an atheromatous stenosis within the anastomotic systems may, in conjunction with a drop in blood pressure, produce an infarct of this region, the region forming a "last field" according to the interpretation of Zülch, 1953—1954 and Schneider, 1966. Secondly, the p.t.o. triangle lies within the axis of the privileged pathway for emboli of the middle cerebral artery. Here we have to emphasize the recent papers of Castaigne et al. (1965) in which they reported that the cause for certain c.v.a. could be emboli from blood platelets and red blood corpuscles which arise on arteriosclerotic plaques of the carotid or middle cerebral arteries.

It has therefore been particularly interesting for us to take up our studies of this p.t.o. region again after having been influenced by some recent observations and to try to define the clinical and electrical features, their etiology, the anatomo-pathology and pathophysiology of this syndrome of the p.t.o. triangle. Eighteen personal cases were available for this investigation.

Material Used for this Study

Eight of the patients were male and 10 female. Their age at the beginning of their symptoms was between 43 and 85 years. Three were over 80, 6 between 70 and 80, 8 between 60 and 70 and only one patient was less than 50, i.e. 43 years (patient Ven.).

The patients can be divided into two groups, in 15 the vascular accident was the only cause of the neurological symptoms, while in 3 another cerebral lesion was superimposed and might also have played a certain role (meningeal hemorrhage, capsular hemorrhage, meningo-encephalitic process).

Etiology

There was a history of important neurological symptoms in 4 patients. Two had recurrent transient episodes. In the other two cases there were single episodes. They had a duration of months or years (3 patients) or only of a few hours (patient Leg.). They consisted of an aphasic, confusional, hemiparetic syndrome or of amblyopia.

Fourteen of the 18 patients had, at the same time, one or several other associated lesions. There were 5 patients with hypertension, 5 others with a cardiac lesion (auricular fibrillation, cardiac insufficiency), 4 with renal insufficiency (functional or organic), 3 with diabetes, 3 with a progressive neurological disorder, 1 with intense dehydration, 1 with porto-caval encephalopathy, one with cardio-vascular collapse, and one each with anaemia, embolism in the extremities, syphilis and recent cranial trauma.

In 9 patients several of the following clinical complications acted together such as: fibrillation, embolism of the extremities, dehydration and uraemia in patient Tin., cardio-vascular collapse, respiratory insufficiency in patient Lat., hypertension and anaemia in patient Bar., hypertension and serologically positive syphilis in patient Mas., hypertension, cardiac insufficiency, old renal tuberculosis in patient Mat., cardiac insufficiency and diabetes in patient Uz., hypertension and auricular fibrillation in patient Dah., uraemia, diabetes and intracranial hematoma in patient Lom., and finally hypertension and uraemia in patient Ser.

Initial Symptoms

The beginning of the symptoms was variable: in 11 cases the onset of the initial symptoms was very sudden as in a c.v.a. and led to the patient's admission. In 4 other cases the p.t.o. syndrome appeared very suddenly but during the course of another affection such as arterial embolism in a cardiac patient or diabetic coma, heavy respiratory insufficiency or hepatic coma.

In 3 patients the p.t.o. syndrome developed during a neurological affection and only in one patient was the syndrome slowly but steadily progressive (obnubilation).

The initial symptoms were as follows: isolated disorders of consciousness (mental disturbance or loss of attentiveness) without any symptoms of irritation or deficit in 6 patients; disorders of consciousness together with neurological deficit (hemiplegia, aphasia) in 6 patients and isolated aphasia in 1 patient.

State on Admission

The state was different in the various patients. Symptoms in the sphere of consciousness and other more specific symptoms of the p.t.o. region were prevalent but the patients were often difficult to examine because of their psychological state.

There were disorders of consciousness in 15 patients, which could be a light obnubilation during the whole time (patient Bat.) or progress to akinetic mutism. The latter was often only a transition to actual coma and death. Eleven patients showed signs of mental confusion during the development of the syndrome. Signs of "excitation" were found in 10 patients of whom 4 showed partial or total hemiclonus and 2 hemiclonus followed or interrupted by general seizures. There were 2 patients with generalised fits, 1 patient with Jacksonian fits starting in the arm, 1 patient with temporal lobe seizures and another with visual hallucinations.

The epileptic fits were always multiple and in 3 cases they progressed to status epilepticus.

The neurological deficits consisted of 8 cases with motor left sided hemiplegia, 6 cases with aphasia, 2 cases with hemianopia and 1 case with cortical blindness.

Development of the Clinical Signs

The development of clinical signs was of ill-omen in the majority of cases: there were 11 deaths, but 5 cases had a favorable outcome; 2 patients were lost to follow-up.

Deterioration and death occurred rapidly in 6 cases (in 2 cases the patients died before the third day, in four cases before the seventh day). In another 5 patients there were periods of transient improvement followed by deterioration and death after the first week.

It is interesting to note that in some cases the deterioration in neurological status ran parallel with the signs of vascular deficiency.

In the patient Tin., the neurological symptoms appeared when there was a cardiac insufficiency with auricular fibrillation and peripheral embolism. In the case Bla., the neurological signs also followed cardiac insufficiency. In the patients Mat. and Gir., the appearance or deterioration of the neurological signs followed a drop of blood pressure from 220 to 150 systolic pressure and 180 to 140, respectively. In the case of Lat. where there was an acute cor pulmonale with respiratory insufficiency similar falls in pressure were observed. Finally, in the patient Dah., the neurological deterioration ran parallel with the development of extreme dehydration.

The Electroencephalogram

The most prominent EEG signs included a general slowing of the basic rhythms usually more marked in one hemisphere. This hemisphere was then the site of far more characteristic signs but these could escape our attention because of their transient nature. Slow or sharp, repetitive waves occurred more or less periodically at about 1 c/s intervals. They were diffusely spread over the hemisphere but had a predilection for the p.t.o. region, where they could remain localized. In some cases this predilection was not so strictly localized and they were found in the high rolandic region near the vertex or even more frontally in the fronto-medial region. Occasionally these sharp waves were bilateral and even more marked in the ischemic hemisphere where the slow waves were maximal. In other cases, before the periodic activity turned up or together with it, rhythmical discharges occurred whose frequency progressively got slower and then assumed the characteristic aspect of a critical epileptic activity. It should be noted that they were similar to the activity which accompanied epileptic fits where they occurred. It is essential to mention that in our 18 cases partial localized myoclonus was not detected clinically, apparently because it was very mild and confined to the proximal segment of one extremity. Moreover, it was often transient, as were the periodic sharp waves which accompanied it and which made its recognition difficult. In some other cases (not used for this paper), they were well defined and had a latency of 70 msec (GASTAUT, 1968).

Pathology

An autopsy was carried out in 4 cases.

Topography of the lesions: The lesions were most marked in the p.t.o. region but often they were also more diffuse. In the case Ser., they were located on the right of the mesial aspect of the occipital lobe, while on the left they were located for the greatest part in the upper parietal lobe and in the 2nd and 3rd occipital convolutions. In the case Mic., they were older and situated more in the right temporal lobe; on the left, they were more in the p.t.o. region and in the inferior-posterior parts of this region. In the case Lat., the lesions were found particularly in the posterior regions, in the right p.t.o. triangle and at the left occipital pole. Finally, in the case Bar., they were situated in the left parieto-occipital region.

There were also more diffusely located lesions.

Multiple infarcts: In the case Mic., infarcts were visible at the first left frontal convolution near the pole, on the right centre median of LUYS and in the left putamen.

In the case Ser., they were on the left in the ascending frontal convolution and inferior parietal region, in the sylvian fissure, in the depths of both superior frontal convolutions, in both ascending frontal convolutions, in the interparietal circonvolutions, and in the outer third of the caudo-putaminal association fibers.

Sometimes localized lesions were found intermingled with diffuse alterations of the "chronic ischemic cell change" type (Bar., Lat., Mic.). It was interesting to see, in the cases Lat. and Miv., marked changes occurred in Sommer's sector, and in the granular and particularly the Purkinje cell layer of the cerebellum, i.e. in zones which are particularly sensitive to ischemia.

Macroscopic appearance: Macroscopic lesions were usually of the anoxic/ischemic type. They were most prominent in the 3rd and 5th cortical layers which are known to be most sensitive and sometimes (case Lat.) had a characteristic laminar appearance. Naturally, the morphology varied with the development and age of the process: edematous imbibition, recent status spongiosus with neuronal changes of the ischemic type (Lat.) or a more advanced necrosis with astrocytic glial proliferation or cystic transformation with capillary proliferation and numerous scavenger cells partly filled with hematein pigment (case Mic.). Often changes of different ages were superimposed in the same brain.

Vascular lesions: These were usually diffuse and found around the basal vessels (Bar., Mic., Lat.) and around arterioles of middle and small caliber.

Summary of the Syndrome

This syndrome occurs almost exclusively in elderly persons, mostly arteriosclerotic, who have suffered for a shorter or longer time from chronic circulatory insufficiency. It starts either very abruptly or during an intervening affliction and particularly when the latter is of a cardiovascular nature.

An electro-clinical syndrome of a very particular nature is observed: from the clinical standpoint disorders of consciousness of various degree may occur. These may increase up to a confusional state. A neurological deficit and epileptic and myoclonic fits may often be seen. They are usually only partial, follow discharges at the p.t.o. region and can end in a status epilepticus. The course is variable, may be serious, and is often influenced by the cardio-vascular state and the systemic arterial pressure.

From the point of the EEG, a general slowing of the background rhythms is seen, associated—during a time course of some hours up to several days—with periodic sharp waves with a repetition of 1 c/s. These sharp waves are diffuse but have a predilection site in the p.t.o. region where they may be mainly localized. They are mostly asymmetric and may be limited to one hemisphere. Moreover, apart from these sharp waves of a low frequency, one can see true EEG seizures or even electro-clinical fits where there are discharges commencing with rapid frequency and later slowing down.

From the anatomical standpoint, the lesions most closely resemble "spongionecrosis". They are rather diffuse but may have a cortical predilection. Their preferential site is the parieto-temporo-occipital region.

Comments

The syndrome which we have described has been named by two of us (Gastaut and Naquet, 1965) the "diffuse cortical ischemic syndrome with an extraterritorial (borderline) predilection".

This makes some anatomical, clinical, electro-clinical and physiopathological comments necessary.

1. Anatomical: This is a predominantly "extraterritorial" diffuse ischemic syndrome. It seems necessary to emphasize the diffuseness and cortical site where a spongy ischemic state is produced. The lesions never occupy a single arterial territory. But, they are predominantly seen in the p.t.o. cortex. There are chronic ischemic alterations of the neurons and even multiple softenings often of a different age. These findings suggest a circulatory subacute insufficiency affecting a brain already damaged by diffuse arteriosclerosis. Lesions in Sommer's sector, in the Purkinje cells of the cerebellum and in the 3rd and 5th layer of the cortex resemble those seen in cases of cardiac arrest.

2. Clinical: Signs of focal and diffuse cerebral lesions are simultaneously present. In less than 50% there are motor deficits; likewise, visual symptoms (hemianopia, cortical blindness, visual hallucinations) and disorders of symbolic expression are occasionally seen. These are all difficult to assess because of the disturbances of consciousness. They are, however, easily explained by the lesion of the p.t.o. triangle.

The intellectual disorders, and the changes in consciousness—which may end in actual coma—are probably referable to the diffuse cortical lesions and only secondarily to brain stem lesions.

3. Electro-clinical: This diffuse "extraterritorial" (borderline) syndrome is characterized by the presence of epileptic phenomena either of a purely electrical nature or of electro-clinical origin. The EEGs are often characterized by the presence of periodic or pseudo-periodic sharp waves predominantly in the p.t.o. region. The clinical manifestations are dominated by numerous focal clonic fits which sometimes end in status epilepticus. There is often myoclonus strictly localized to one part of half the body; it may escape clinical observation because of its slight intensity.

These excitatory disorders make two comments necessary:

a) they follow an ischemic anoxia,

b) experimentation (GLASER and SJAARDEMA, 1946; HARVEY and RASMUSSEN, 1951; MEYER, 1956; BICKFORD, 1958; MORRELL and FLORENZ, 1958; GASTAUT et al., 1958; NAQUET and FERNANDEZ-GUARDIOLA, 1960) shows that ischemic anoxia can provoke excitatory discharges.

Other authors have reported numerous cases in which paroxysmal phenomena appeared in vascular lesions which were examined clinically and morphologically (BAROLIN et al., 1962; CHATRIAN et al., 1965).

On the other hand, these epileptic manifestations, with their clinical picture, topography and electrical phenomena, raise certain considerations. NAQUET (1965), NAQUET, FRANCK and VIGOUROUX (1965), GASTAUT and NAQUET (1965) have noticed, in cases of chronic vascular cerebral insufficiency in the aged, electrical subclinical discharges which were rarely of a synchronous and symmetrical nature, were more or less generalized but always predominantly localized in the p.t.o. region.

NAQUET et al. (1965, 1966) have introduced air emboli into the carotid artery in cats and monkeys. They found: first a general, then a local arrest of circulation in the arterioles, mainly in the distal segments. This was followed by corresponding electro-clinical signs. After a phase of electric depression there was regeneration, then irritative discharges which often exaggerated into status epilepticus, or resembled periodic sharp waves, more or less generalized but with a predominance in the p.t.o. region. Anatomically, the lesions resembled spongy necrosis with a predilection for the posterior regions of the brain.

These statements correspond to the human pathology, as for instance the air embolism in cardiac surgery where one finds the same irritative phenomena and the same predilection for the posterior regions (ARFEL et al., 1967). They are responsible for the periodic discharges after cardiac arrest (HOCKADAY et al., 1965) or the subacute presenile encephalopathy of JONES-NEVIN, a syndrome which by most authors is placed near CREUTZFELD-JAKOB's disease. It is worth emphasizing that this diagnosis was actually discussed in regard to one of our patients, who had in fact a cortical blindness (HEIDENHAIN's disease).

Table 1.

Cases	Previous neurolog. accidents	Associated lesions	First sign
M. Tin. 82 years	+	Auricular fibrillation, arterial embolism into extremities, dehydration, hyperazotemia	Mental confusion
M. Lat. 65 years	+	Chronic respiratory insufficiency, atelectasias, cardiovascular collapse	Clonic hemiseizures
M. Ven. 43 years	Mental confusion with transient aphasia	+	Mental confusion
M. Bar. 71 years	+	Arterial hypertension, anemia of unknown origin	Right hemiplegia
M. Mich. 79 years	Transient hemipareses 3 times	+	Leftsided fits
M. Gir. 73 years	+	Diabetic coma and attacks of hypotension	Bravais-Jacksonian fits starting in the arm
M. Mass. 60 years	+	Hypertension Wassermann-reaction of blood	Aphasia
M. Bat. 68 years	+	Cerebral trauma, meningeal hemorrhage	Grand-Mal seizures
M. Bla. 85 years	+	Auricular fibrillation with cardiac insufficency	Loss of consciousness
M. Mat. 70 years	+	Hypertension, renal tuberculosis, left cardiac insufficiency	Right hemiplegia with aphasia and hemianopia
M. Uz. 83 years	+	Auricular fibrillation	Somnolence then mental confusion
M. Dah. 78 years	+	Hypertension, auricular fibrillation	Rightsided hemiplegia with aphasia
M. Leg. 69 years	Transient aphasia	+	Rightsided clonic fits, mental confusion
M. Lom. 66 years	+	Diabetic hyperazotemia hemorrhage in the right parieto-temporo-occipital trigonum, operated	Leftsided hemiplegia
M. Ser. 60 years	2 episodes of hemiparesis, 1 of amblyopia	Hypertension, Hyperazotemia	Left hemiplegia
M. Lan. 70 years	+	Meningo-encephalitis	Mental confusion
M. Dil. 60 years	+	Alcoholism and portocaval syndrome	Leftsided hemiclonic fits
M. P. 64 years	+	+	Aphasia

Table 1. (continued)

Focal signs	Consciousness	Convulsions	Evolution
+	Coma	+	Death 2nd day
Right Hemiparesis	Coma	O	Death 5th day
Aphasia, optic hallucinations	Mental confusion	Temporal fits	Favorable
Right hemiplegia	Coma	Hemilateral fits	Death 18th day
Left hemiplegia	Coma	O	Death 2nd day
+	Coma	O	Death 8th day
Aphasia	+	+	Favorable
Leftsided hemiparesis	Somnolence, mental confusion	O	Favorable
+	Mental confusion later coma	+	Death 21st day
Right hemiplegia with aphasia and hemianopia	Mental confusion	+	Death 7th day
+	Coma	+	Death 6th day
Rightsided hemiplegia with aphasia	Mental confusion, later coma	+	Death 12th day
Somnolence	Somnolence	Grand-Mal seizures	Favorable
Left hemiplegia, homonymous hemianopia	Mental confusion, later coma	+	Death 16th day
Left hemiplegia, cortical amaurosis	Mental confusion, later coma	Grand-Mal seizures	Death 5th day
+	Coma	Rightsided clonic fits	
+	Coma	General convulsions	
Aphasia	+	+	Favorable

4. Pathophysiological: The p.t.o. "triangle" syndrome may be explained by a circulatory deficiency. A drop of blood pressure of whatever origin (hypotension, cardiac failure) may provoke irreversible cerebral lesions. These lesions will be localised at the p.t.o. triangle because this part is the farthest away from the origin of the arterial trunks and also because hypotension may endanger the function of the anastomoses Meyer and Denny-Brown, 1957). The appearance of such a syndrome will be favoured by preexisting arteriosclerosis of the extra- or the intracranial vessels which per se diminishes the cerebral blood flow. This hypothesis may be supported by etiological and anatomical arguments.

Etiological: It was surprising to find that in 5 patients the development of neurological symptoms was in parallel with the hemodynamic situation: attacks of cardiac insufficiency, drop of blood pressure, intense dehydration. In some of our cases the symptomatology followed myocardial insufficiency with decreasing systemic pressure.

It is reasonable to suggest that these hemodynamic factors associated with diffuse arteriosclerosis give rise to a decrease in cerebral blood flow severe enough to provoke irreversible lesions. In some cases other factors were operating to diminish the supply of cellular oxygen: for instance anoxia from respiratory insufficiency or anemia.

Anatomical: The site of the lesions (diffuse lesions predominantly in the p.t.o. region), the absence of thromboses in the main intracranial arteries, and the changes found particularly in those parts of the parenchyma which are susceptible to ischemia also favour the above formulated hypothesis.

Is this to say that all our cases follow only one pathophysiological pattern and that there are not cases of cardiac embolism or else emboli from the carotid or middle cerebral artery which could lead to such a p.t.o. syndrome? Anatomically the absence of such emboli does not speak against this possibility since it is well known that they become resolved very quickly (Alajouanine *et al.*). Moreover, such vessels can be recanalised. The multiplicity of the symptoms could be explained by the multiple lesions or could be secondary to hemodynamic changes which follow the primary infarct.

With regards to etiology, most of our patients showed a cardiac insufficiency and disturbances of rhythm which favour, eminently, the formation of thrombi and their embolization: in one of our cases the neurologic symptoms followed soon after an embolism of the lower extremities.

Finally, the p.t.o. region is in the terminal territory of the middle cerebral artery, a favoured route of emboli; one of us has (Vigouroux, 1966) in a thesis devoted to air embolism in the carotid (a lesion where the alterations in the p.t.o. region are predominant), discussed the above mentioned hypothesis and recently Lhermitte *et al.*, 1968, had the occasion to emphasize the frequency of an embolic origin of the sylvian thromboses.

Therefore, at this moment an authentic theory cannot be yet formulated. However, in the light of some of the cases which we have had the opportunity of examining in detail and describing above where the neurologic symptomatology followed hypotension so strictly, the drop of circulation may have acted as a primary causative agent. This hemodynamic theory seems most acceptable especially as the same anatomical changes have been caused by cardiac arrest.

However, the two hypotheses are not entirely contradictory. Whichever may be the primum movens of the neurologic sequelae (a hemodynamic or embolic process), the reestablishment of the circulation in the ischemic region, the triggering of the supplementary anastomotic circulation, and finally the sequel, multiple manifestations or the limitation of the softenings, remain functionally dependent on the particular hemodynamics of this region.

References

ADAMS, R. D., VAN DER EECKEN, H. M.: Vascular diseases of the brain. Ann. Rev. Med. 4, 213—252 (1953).

ALAJOUANINE, TH., BERTRAND, I., CASTAIGNE, P., LHERMITTE, F., CAMBIER, J., GAUTIER, J. C.: Les nécroses ischémiques dans les obstructions bilatérales de la carotide interne. A propos de trois observations anatomo-cliniques. Rev. neurol. 102, 113—129 (1960).

— CASTAIGNE, P., LHERMITTE, F., CAMBIER, J., GAUTIER, J. C.: Les obstructions bilatérales de la carotide interne. Sem. Hop. Paris 35, 1149—1160 (1959).

— — — CLAY, R.: Aspects artériographiques des thromboses artérielles cérébrales et de la circulation de suppléance. Sem. Hóp. Paris 33, 2135—2150 (1957).

— — — GAUTIER, J. C.: Les anastomoses des artères cérébrales, leur rôle de suppléance. Sem. Hóp. Paris 35, 1135—1141 (1959).

ARFEL, G., CASANOVA, C., NAQUET, R., PASSELECQ, J., DUBOST, CH.: Etude électro-clinique de l'embolie gazeuse cérébrale en chirurgie cardiaque. Electroenceph. clin. Neurophysiol. 23, 101—122 (1967).

BAKER, A. B., DAHL, E., SANDLER, S.: Cerebrovascular disease. Etiologic factors in cerebral infarction. Neurology (Minneap.) 13, 445—454 (1963).

BAROLIN, C., SCHERZER, E., NAQUET, R., GASTAUT, H.: Etude électro-clinique des états de mal épileptique survenant chez les apoplectiques. Rev. neurol. 107, 242—243 (1962).

BICKFORD, R. G.: Discussion of MORRELL and FLORENZ paper "Modification of the freezing technique for producing experimental epileptogenic lesions". Electroenceph. clin. Neurophysiol. 10, 186 (1958).

CASTAIGNE, P., LHERMITTE, R., GAUTIER, J. C.: Rôle des lésions artérielles dans les accidents ischémiques cérébraux de l'athérosclérose. Rev. neurol. 113, 5—32 (1965).

CHATRIAN, G. E., SHAW, G. M., LEFFMAN, H.: The significance of periodic lateralized epileptiform discharges: an electrographic, clinical and pathological study. Electroenceph. clin. Neurophysiol. 17, 177—193 (1964).

FISCHER, C. M.: Cerebro-vascular disease. Physiopathology, diagnosis and treatment. J. chron. Dis. 8, 419—447 (1958).

GASTAUT, H., NAQUET, R., REGIS, H., SALAMON, G.: Sur les effets électrographiques d'une anoxie cérébrale de longue durée. C. R. Soc. Biol. (Paris) 152, 1251—1253 (1958).

— Séméiologie des myoclonies et nosologie analytique des syndromes myocloniques. Rev. neurol. 119, 1, 1—30 (1968).

— NAQUET, R.: Etude électroencéphalographique de l'insuffisance circulatoire cérébrale. Symposium international sur la circulation cérébrale, p. 163—191. Paris: Sandoz 1965.

GLASER, H. A., SJAARDEMA, H. H.: Effect on the electroencephalogram of localized pressure on the brain. J. Neurophysiol. 9, 63—72 (1946).

HARVEY, J., RASMUSSEN, TH.: Electroencephalographic changes associated with experimental temporary focal cerebral anemia. Electroenceph. clin. Neurophysiol. 3, 341—351 (1951).

HOCKADAY, J. M., POTTS, F., EPSTEIN, E., BONAZZI, A., SCHWAB, R. S.: Electroencephalographic changes in acute cerebral anoxia from cardiac or respiratory arrest. Electroenceph. clin. Neurophysiol. 18, 575—586 (1965).

LHERMITTE, F., GAUTIER, J. C., DEROUESNE, C., GUIRAUD, B.: Ischemic accidents in the middle cerebral artery territory. A study of the causes in 122 cases. Arch. Neurol. (Chic.) 19, 248—256 (1968).

MEYER, J. S.: Studies of cerebral circulation in brain injury. Cerebral contusion, laceration and brain stem injury. Electroenceph. clin. Neurophysiol. 8, 107—116 (1956).

— DENNY-BROWN, D.: The cerebral collateral circulation. Factors influencing collateral blood flow. Neurology (Minneap.) 7, 447—458 (1957).

Morrell, F., Florenz, A.: Modification of the freezing technique for producing experimental epileptogenic lesions. Electroenceph. clin. Neurophysiol. 10, 187—188 (1958).

Naquet, R.: L'importance du carrefour parieto-temporo-occipital dans la pathologie épileptique d'origine vasculaire. Brux. méd. 45, 378—384 (1965).

— Arfel, G., Choux, M., Dubois, D.: Etude expérimentale de l'embolie gazeuse par voie carotidienne chez le chat. Electroenceph. clin. Neurophysiol. 20, 181—196 (1966).

— Fernandez-Guardiola, A.: Effets de différents types d'anoxie sur l'activité électrographique cérébrale spontanée et évoquée chez le chat. J. Physiol. (Paris) 52, 885—902 (1960).

— Franck, G., Vigouroux, R.: Données nouvelles sur certaines décharges paroxystiques du carrefour parieto-temporo-occipital rencontrées chez l'homme. Zbl. Neurochir. 25, 153—180 (1965).

— Vigouroux, R. P.: Embolies expérimentales par voie carotidienne chez le babouin (Papio papio). Rev. neurol. 114, 339—360 (1966).

Schneider, M.: Hypoxie and Anoxie. Therapiewoche 6, 217—221 (1966).

Vigouroux, R. A.: L'embolie gazeuse cérébrale expérimentale par voie carotidienne. Thèse Méd., Marseille 1966, 117 p.

Zülch, K. J.: Neue Befunde und Deutungen aus der Gefäßpathologie des Hirns und Rückenmarks. Zbl. allg. Path. path. Anat. 90, 402 (1953).

— Mangeldurchblutung an der Grenzzone zweier Gefäßgebiete als Ursache bisher ungeklärter Rückenmarksschädigungen. Dtsch. Z. Nervenheilk. 172, 81—101 (1954).

— Behrend, R. C. H.: The pathogenesis and topography of anoxia, hypoxia and ischemia of the brain in man. In: Cerebral anoxia and the electroencephalogram, p. 144—163. (Gastaut, H., Meyer, J. S., Eds.). Springfield (Ill.): Charles C. Thomas 1961.

Research on Cerebral Circulation
at the Faculty of Medicine of Toulouse*

G. Lazorthes, J. Espagno, A. Bes, Y. Lazorthes, and L. Arbus

The problems of cerebral circulation in health and disease have occupied the center of research at the neurological and neurosurgical clinics of the University Hospital of Toulouse for a long time.

Since 1925, Riser has devoted his research activities to physiological and pharmacological problems of the cerebral circulation and has reported on this at International Meetings (1935, 1936). The results of his research have provided us with basic data for our present knowledge. This is especially true since the autonomous position of the cerebral vasomotor system has been proved and thus the "autoregulation" of the cerebral blood flow is becoming progressively clearer even now.

From 1939 to 1949 G. Lazorthes studied the innervation of the vessels, particularly that of the cerebral arterial supply (1949). From then on, with the aid of his collaborators, he made further investigations on the various vascular segments i.e. the cerebral veins (1953), the arteries of the cerebellum (1955) and of the diencephalon (1956), then the anterior cerebral artery (1957), the spinal arteries (1957) and arteries of the brain stem (1958) and the cerebrum (1960). These investigations were integrated into one volume (1961): "The vasculature and the cerebral circulation" in collaboration with L. Campan. Finally in the last years, our studies were directed towards the auxiliary supply by anastomoses in the cerebrum and the spinal cord (1968).

In 1952 J. Espagno oriented towards the hemodynamics of the cerebral circulation. He published a thesis on the (basic) cerebral blood flow in which he demonstrated for the first time in Europe that the autoregulation of the cerebral blood flow was remarkably constant even when a level of arterial pressure of 70 to 80 mm Hg was reached.

Then after 1960 J. Geraud and A. Bes et al. developed a laboratory for the study of the hemodynamics and metabolism of the brain. Meanwhile, they have introduced Krypton 85 for the global measurement of the cerebral circulation. With this method they have studied the variations of the circulation in c.v.i. and in the state of dementia (M. Delpla, 1963; J. P. Marc-Vergnes, 1965). In 1964, we introduced the Xenon 133 method of Lassen and Ingvar which is administered via carotid injection and with collimation in one or several channels. In 1966 we used, as a routine method (Mallet and Weall), the inhalation of Xenon. Since then we have been particularly eager to study cases of severe coma and have tried to investigate systematically the cerebral metabolism of such cases.

---

* From the Centre Hospitalier and Universitaire de Toulouse, Hôpital de Purpan, Toulouse.

Finally, in 1964 I. Espagno and Y. Lazorthes introduced a new method of measuring cerebral blood flow in man by locally injecting microdoses of Xenon 133 and following its clearance curve during an open skull operation. Our first results were published at the II. Lund Symposion in 1965 and our total results were collected in the thesis of Y. Lazorthes.

Since then we have completed our technique of local cerebral blood flow measurements by microprobing the cortical vessels and studying both the global and the local metabolism. Recently, the possibility of registering rapid changes by means of thermo probes, as suggested by Betz and Wüllenweber, has improved our techniques of hemodynamic studies.

The principal aspect of our method is that we are able to measure the perifocal lesions and changes in the cerebral parenchyma with regards to their hemodynamics and metabolism.

In the following account we shall limit the description of our results to those data which one can either consider as physiological or pathological and we shall refrain entirely from all experimental, pharmacological problems.

I. Normal Results

Measuring the cerebral blood flow by intraparenchymatous injection of Xenon 133 has allowed us:

1. to confirm the important contrast between the blood flow of a) the white substance which has a constantly low value of 16 ml/100 g/min and b) the cortex which has a variable and higher value of 85 ml/100 g/min whereas the global average amounts to 50 ml/100 g/min. These data are in good agreement with those given by such other authors as Lassen, Ingvar and Fieschi who based their data on the intracarotid injection of Xenon 133;

2. to emphasize the heterogeneous character of the cortical demand by showing that there are in the cortex two different parts, one rapidly perfused (f_1 = 85 ml/100 g/min) and one which is only slowly perfused (f_2 = 12.5 ml/100 g/min). The latter are values, which according to the "external" methods of measurement, resemble those of the white substance;

3. to define the existence of significantly different values of local blood flow between different cortical territories without these areas being cytoarchitectonically related. At this point, we can only say that the parietal and occipital lobes have the highest blood demand. It has appeared to us to be impossible or even to be an illusion to hope to define a "map" of the various cortical demands that relate to the different "functional" areas. This is because of the different types of persons examined and because of the variability and the minimal "physiological" conditions of the measurements. Therefore we have oriented our research entirely towards the field of pathology.

II. Pathological Results

We have measured the blood demands in pathological territories in operated patients who possessed localized lesions either of brain tumors vascular nature.

A. Vascular Cerebral Lesions

The data of J. Geraud, A. Bes and M. Delpla concerning the measurements of the global blood flow in the c.v.i. have allowed us to distinguish three different stages of development:

In the first (V 1) which corresponds to the clinical stage of the beginning of the cerebral vascular lesion the usual demand corresponds to a "paranormal" value of

r. T.B.F.
0,2 mC 133 Xe
(0,2 – 0,3 ml)
tumoral artery

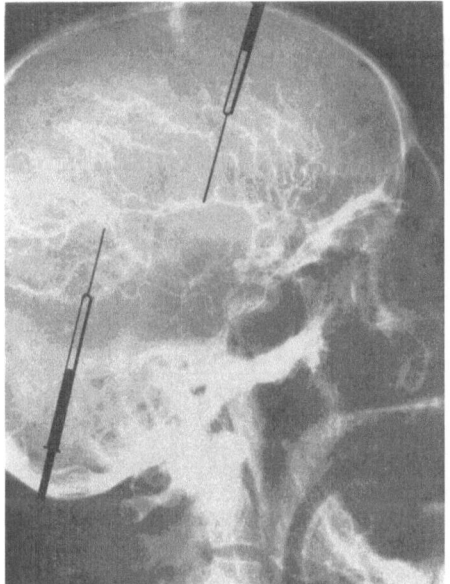 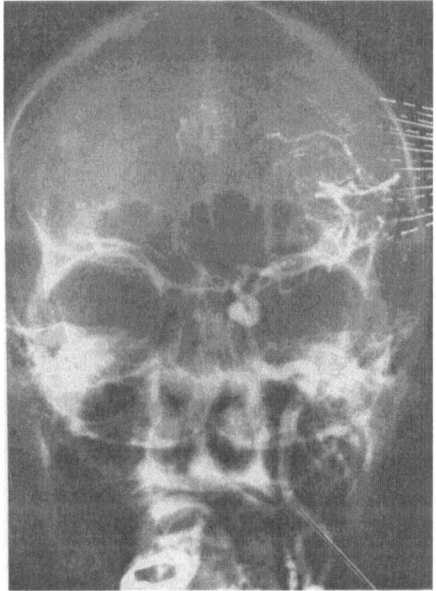

I. T.B.F
0,1 mC 133 Xe (1-5ul)
intraparenchymatous
injection

r. C.B.F.
0,5 mC 133 Xe (5 ml)
internal carotid artery

Fig. 1. Methods for the clearance evaluation of Xenon 133: Injection of the isotope into the carotid r CBF; injection of the isotope into the tumor pedicle r TBF; injection into the tumor l TBF

45.8 ml/100 g/min. In the second (V 2) group of more severe vascular accidents the values decrease to 36.2 ml/100 g/min average; here some patients still have a normal oxygen consumption, whereas it has already decreased in the very severe cases. In the third group (V 3), which corresponds to vascular patients with corresponding mental disturbances, the demand is down to 29.9 ml/100 g/min.

The use of multiple collimators has allowed Lassen and Ingvar to show modifications of the *local* demand near a pathological cerebral territory. Moreover, Lassen described in the first group in the phase of an acute reaction a syndrome of "luxury perfusion" with transient increase of the cerebral blood demand in the zone of the lesion which was induced by a secondary hyperemia due to local metabolic acidosis.

Contrary to this, in the second type of lesion this luxury perfusion syndrome disappears and gives way to a regional decrease of the CBF.

We have been particularly eager to study the local disturbances of the CBF in the ring-like zone of the cerebral parenchyma around the spontaneous intracerebral mass hemorrhages. We have always found a local CBF which was very diminished, whether it is the cortex (40 ml/100 g/min) or the white substance (10 ml/100 g/min). This confirms the existence of an ischemic zone around the hemorrhage. This ring-

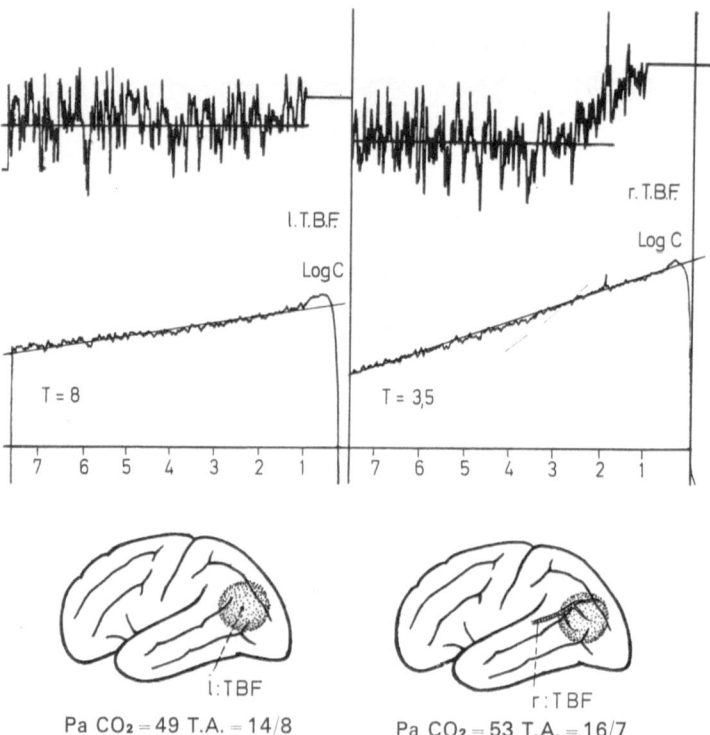

Fig. 2. Blood flow of a glioblastoma. On the left: after an intratumoral injection of Xenon 133; on the right: after injection into the tumor pedicle

like zone is limited because within a certain distance the CBF is normal again. Never have we been able to register a "luxury perfusion". Perhaps because our measurements were not made at or soon after the vascular accident, since we usually operate on our patients on the 4th or 5th day which means that we see our patients later, i.e. at the second (V 2) stage.

B. Brain Tumors

1. Methods of Examination

From the point of hemodynamics we have up to now examined not only the circulation in 21 cases of gliomas, metastases or meningeal tumors, but we have also, in 28 cases, measured their influence on the peritumoral parenchyma. The CBF has been studied by the clearance method with Xenon 133 which we have injected in three different ways:

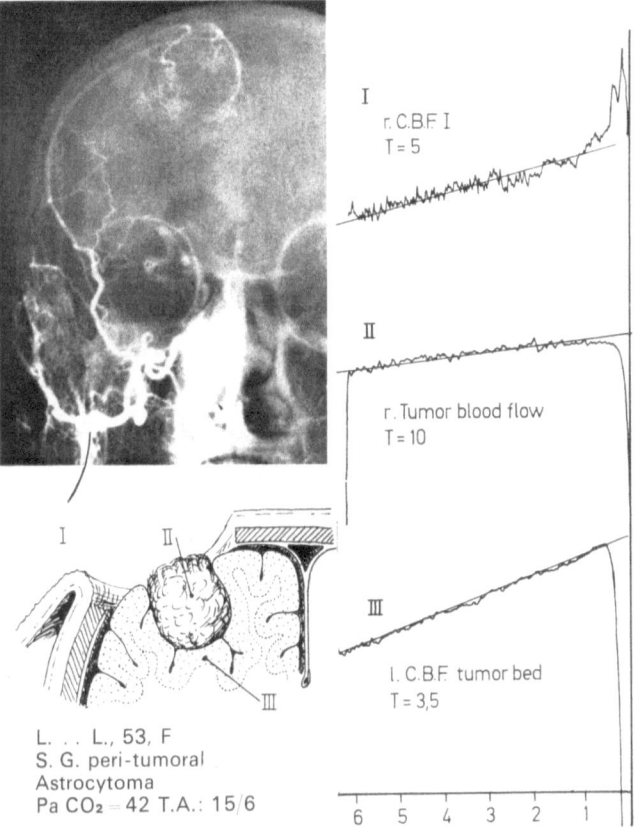

Fig. 3. Hemodynamic study of a meningeoma on the convexity of the skull. Cerebral blood flow: I after injection of Xenon 133 into the external carotid; II after injection into the meningeoma itself; III of the meningeoma bed after extirpation

a) directly into the tumor,
b) into the arterial branch supplying mainly the tumor,
c) into the internal carotid with registration with either the skull open or closed (Fig. 1).

Only with the two first technics can we strictly evaluate the blood flow of the tumor because by intracarotid injection of the Xenon 133 we measure two blood flows, that of the tumor and that of the neighbouring parenchyma, and this is only possible if the distance of the collimators is very short.

Therefore, we have developed a technique of studying the metabolism by micro-samples of the blood which drains from the tumor and are able to obtain a micro-estimation of:

1. the total volume of O_2 and CO_2,

2. the PCO_2 and pH and of

3. the glucose, lactic acid and pyruvic acid.

Thus, we can compare the local metabolism of the tumor and the global cerebral metabolism measured at the same time [blood is taken from the internal jugular vein; this in contrast to arterial blood taken from elsewhere (carotid, femoral)].

2. Results

a) *Tumorous Blood Flow* (TBF)

Actually it is not possible to formulate quantitative data on the demand of the tumor tissue because one does not know the value of the dilution coefficient of

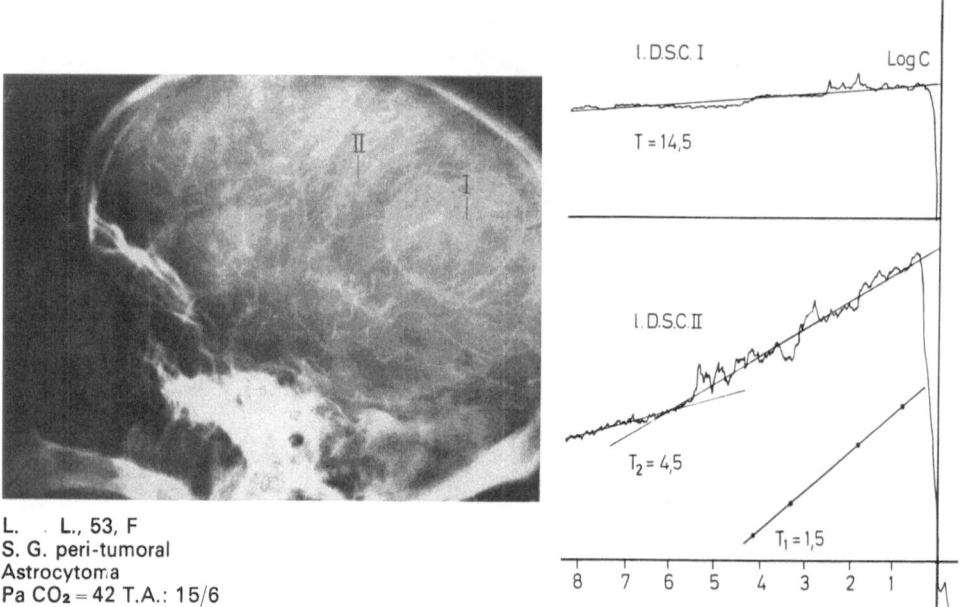

L.   L., 53, F
S. G. peri-tumoral
Astrocytoma
Pa CO_2 = 42 T.A.: 15/6

Fig. 4. The peritumorous edema: CBF I in the peritumoral region; CBF II within the surrounding area

blood-tumor tissue of the Xenon 133. One can therefore only get a "circulation velocity index" which corresponds to the "slope" of the elimination.

Gliomas. The clearance curves of the glial tumors are usually of the multiexponential type which confirms the heterogeneous character of their circulation (Fig. 2). Although these tumors usually show up in the angiograms as hypervascular tumors, their average blood demand is low because their elimination rate is between 8 to 14 min whereas it is 0.8 min for the cortex and 6 to 7 min for the white substance. This is easily understood because our measurements are often taken from necrotic tumor tissue. On the other hand, in some cases we have registered very high values expressive of arteriovenous shunts.

In the field of metabolism we often prove the existence of these arterio-venous shunts by demonstrating the decrease of the a.v. difference in O_2 and in glucose in these so called "red veins".

Meningeomas. We have found a much closer correlation between the rich vascularisation of the tumor (as shown by angiogram and histology) and the rapidity of the "slope" in elimination of the Xenon 133. These may be very variable (from 1 to 10 min) and dependant on the type of the meningeoma (Fig. 3).

b) The Peritumorous Parenchyma

In the peritumoral edema. We have always found a diminution of the CBF of the cortex and white substance in the neighbourhood of the gliomas and metastases. This decrease can even end in a "break down" with a maximum of circulatory arrest. It is perifocal because it is within the bounds of normal CBF. It is, however, associated with a loss of autoregulation (Fig. 4).

In the meningeoma bed. The diminution of the cortical blood flow is due to a mechanism of compression and to a phenomenon known as a vascular "steal" brought about by a detour of the tumor blood. These changes are always maximal in the tumor "bed". They diminish towards the edges and normalize within a distance (Fig. 3). The decrease in the CBF has an important bearing on the prognosis for the recuperation of the neurological deficits.

To conclude this very brief summary of our work, we can say that our investigations are still in their beginning and that we try to orient our studies toward the hemodynamic and metabolic side of the intracranial tumors and then also toward "comas dépassés" for which we try to define the early biological symptomatology.

Morphological Studies in Vivo: Angiography

Arteriography of Collateral Circulation in Cerebrovascular Disease*

W. S. Fields

The efficiency of collateral circulation plays an important part in determining the outcome of stenosis or occlusion in an artery. When a primary vessel is prevented from carrying its contents, others assume this function. Collateral circulation in any part of the body is made possible by the anastomosis of peripheral arteries. Each of the principal arteries has its own peculiar anatomical pattern of anastomotic vessels.

The foundation of arterial circulation lies in vessels which are formed in the early stages of fetal life. Some of these vessels succumb to involution while others are evolving to supply blood to the developing structures. However, anatomical factors are not solely responsible for the development of collateral circulation since functional changes also may influence the effectiveness of the anastomosis.

Neurogenic vasospasm influences considerably the adequacy of collateral circulation in most parts of the body. When vasodilatation is possible, circulation returns more quickly to the part of the body deprived of blood through the primary channel. Removal of sympathetic influences will result in vasodilatation and an increase of blood flow in that part of the body.

In the head and neck the influence of vasomotor mechanisms through the autonomic nervous system is important in the extracranial arterial circulation but less important in the intracranial circulation.

Morphologists have described patterns of anastomotic circulation which are presented meticulously in textbooks of anatomy. However, the introduction of cerebral angiography has provided a new method for visualizing patterns of flow. This has been made possible by refinements in technique, such as serial biplane angiography and cinefluorography. These techniques not only provide information about the anatomical patterns but also assist in understanding the physiological factors which are implicated in the anastomotic circulation, such as velocity and direction of flow.

Collateral circulation becomes exceedingly important in people as they grow older and begin to develop atherosclerotic plaques in various portions of the arterial tree. The most common cause of stenosis in major vessels is atherosclerosis which is usually well localized in specific segments of the arterial circulation. The most common points of narrowing are at the origin of arterial channels or at bifurcations. Before total occlusion occurs, the lumen of the vessel is gradually encroached upon by the

* From the University of Texas, Graduate School for Biomedical Sciences at Houston Neurological Programs, Houston/Texas.

deposition of subintimal atheromata. Eventually the lumen becomes compromised to the point that the flow of blood is so impeded that thrombosis occurs. When this event ensues, the viability of the tissue in the territory normally supplied by the primary vessel depends upon the efficacy of the collateral circulation. Because this process of stenosis is gradual rather than sudden, a more effective collateral circulation develops. Its ability to compensate for the loss of blood supply through the primary channel is thereby enhanced.

In patients suspected of having stenosis or occlusion in the extracranial or intracranial arteries, one should endeavor to accomplish complete visualization of the arterial tree from the great trunks originating in the aortic arch up to and including the intracranial branches. This is an essential prerequisite to proper diagnosis of stenotic and occlusive lesions which may compromise the arterial circulation of the head and neck. Panarteriography should always be undertaken when surgically accessible lesions are encountered and the possibility of employing arterial reconstructive operation to restore blood flow is considered.

No single technique for angiographic studies is superior to all others. In general, the important objective is to utilize a method which will supply sufficient information to arrive at a proper diagnosis and to offer, in addition, as much information as possible regarding prognosis. It is in the latter respect that delineation of collateral circulation becomes exceedingly important.

The extent and adequacy of collateral circulation cannot be fully appreciated when only single films are made. Serial angiography and cinefluorography permit one to follow the bolus of contrast material through both the intracranial and extracranial arterial segments and to visualize the patterns of the flow of blood.

Our preference has been to utilize a technique by which complete visualization of these vessels can be accomplished in one single session. This can be done by studying first the thoracic aorta and its principal trunks by means of retrograde femoral catheterization or a catheter passed through the axillary or subclavian arteries. The thoracic aortogram should be followed by selective catheterization of the individual distal branches in order to visualize each one separately. Aortography frequently produces superimposition of one vessel on another. It is, therefore, necessary to inject each major trunk independently to fully appreciate collateral flow from one part of the circulation into another in the presence of marked stenosis or total occlusion.

Specific patterns of collateral circulation exist for each of the major arteries of the cervical region and in the intracranial cavity. Around the great vessels which originate in the aortic arch, a rich collateral bed compensates for stenosis or complete occlusion in any one of these arterial trunks. Similarly, patterns of collateral are also described for the internal carotid and vertebral arteries which supply the intracranial structures. These patterns of collateral have been identified both from postmortem examination and arteriographic studies.

Morphologic variations on a congenital basis are quite frequent in the circulation of the head and neck, particularly in the circle of Willis at the base of the brain. This circle, a potentially important anastomosis, should provide equitable distribution of blood to all parts of the brain, but it is unreliable as a source of collateral circulation in many individuals. Postmortem studies of the circle of Willis show that the classical pattern described in most anatomical texts is encountered in only 40% of the general population. These vessels when they are anomalous are usually narrow or string-like, but segments of the circle may actually be absent as a result of agenesis or involution during embryonic development. As a result of this variability, the clinical manifestations of occlusion of the vertebral and carotid arteries may vary considerably from one individual to another, and the effectiveness of collateral circulation may be greatly influenced.

Detailed anatomical description of the anastomotic patterns for each of the major vessels has been presented elsewhere and will not be reviewed here. On the other hand, it would be useful to describe the pattern of collateral circulation in internal carotid artery thrombosis, the occlusive lesion most frequently encountered in the extracranial circulation and the one most likely to produce serious neurologic deficit secondary to cerebral ischemia or infarction.

There are no branches in the extracranial portion of the internal carotid artery. Therefore, when this vessel is occluded, the thrombus usually propagates distally to the next point of branching which is at the level of the caroticotympanic artery. On occasion it may pass beyond this branch, particularly if the branch is small and does not carry more than a minute amount of blood back into the main stream. Under such conditions, the thrombosis extends to the ophthalmic artery, the first major branch.

The blood circulates into the distal intracranial portion of the internal carotid artery through the circle of Willis and from external carotid anastomoses. The principal channels are as follows:

1. Across the anterior segment of the circle by way of the anterior cerebral and anterior communicating arteries.

2. Through the ipsilateral posterior communicating artery from the basilar arterial circulation.

3. Through the anterior tympanic branch of the external carotid into the caroticotympanic branch of the internal carotid just below the point where the latter enters the cavernous sinus.

4. Through branches of the internal maxillary artery into the ophthalmic artery resulting in a retrograde flow of blood through the latter. This communication is primarily by way of the recurrent branch of the middle meningeal artery and the lacrimal branch of the ophthalmic artery.

In any of these anastomoses, anomalies may prevent development of an adequate collateral flow. First, when one considers collateral flow from external carotid branches, one must remember that the ophthalmic artery sometimes does not originate from the internal carotid but rather from other external carotid branches. In this circumstance the circulation through the circle of Willis must be sufficient to compensate for the ineffective orbital circulation. On the other hand, anomalies in the circle of Willis in either the anterior or posterior segment may prevent the establishment of compensatory circulation. There is frequently absence or attenuation of the first, or A-1, portion of the anterior cerebral artery between its origin from the internal carotid and the anterior communicating artery. This may occur on either side but is present more often on the right side than on the left. The posterior cerebral arteries in the embryo arise from the carotid system and gradually migrate posteriorly so that post-natally they become branches of the basilar artery. However, quite frequently the primitive pattern persists in adult life. Hence, the posterior cerebral artery on one or both sides originates from the carotid. In such cases there is often an inadequate or absent communication between the basilar and carotid circulations. When one or more of these potential anastomoses has developed anomalously, the carotid occlusion will probably not be well compensated.

Arteriographic studies frequently confirm the fact that complete occlusion of one internal carotid artery does not always result in permanent functional deficit in the brain. Even complete obstruction of both internal carotids or both vertebral arteries may be well compensated by collateral circulation. Marked disability is avoided when the compensatory mechanisms can become operative promptly after the occlusion occurs.

When occlusion occurs in the intracranial arteries, compensatory mechanisms also carry collateral circulation to the territory normally supplied by the primary cerebral vessels. For example, when occlusion develops in the middle cerebral artery at the level of the trifurcation or in one of the branches of the sylvian complex, then collateral circulation takes place over the convexity of the cerebral hemisphere through anastomoses between the leptomeningeal branches of the anterior and posterior cerebral arteries and those of the middle cerebral artery. Similarly, when occlusion takes place in the anterior cerebral or posterior cerebral arteries, collateral circulation is possible over the convexity through the leptomeningeal branches of the other vessels. However, when the obstruction develops in the origin of the main stem of the middle cerebral artery at the level of the intracranial bifurcation of the internal carotid artery, collateral circulation usually is insufficient to compensate for the loss of the entire middle cerebral flow. An occlusion at this site is proximal to the origin of the lenticulostriate arteries. Consequently, infarction occurs in the depths of the brain in the region of the internal capsule. Such a lesion is accompanied by irreversible neurologic deficit.

In the vessels of the posterior fossa, occlusion is also reasonably well compensated if it occurs in certain locations. For instance, when the thrombus is formed in the mid portion of the basilar artery, there is sufficient collateral circulation over the convex surfaces of the cerebellar hemispheres to provide blood to the upper and lower brain stem through communications between the leptomeningeal branches of the posterior inferior cerebellar and anterior inferior cerebellar arteries and the superior cerebellar artery. When the occlusion occurs at the level of the bifurcation of the basilar artery or at the confluence of the vertebral arteries, then it is less well compensated, and brain stem infarction usually results.

It should be evident from this review that effective collateral circulation is possible at each of three levels in the aortocranial circulation. First, in what we might call the proximal extracranial vessels, second, the distal extracranial vessels and, third, the intracranial vessels. Probably because of the size and number of available collateral channels and the proximity to the source of cardiac output, the proximal extracranial lesions are those best compensated when occlusion occurs. Intracranial vascular occlusion as a result of being farthest from the power source are the least well compensated.

Finding occlusion in one vessel without disease in any of the others is a rather rare occurrence. It is obvious that the atherosclerotic process is a generalized one even though it may be segmental in character. As a consequence, multiple lesions are more frequently encountered than single ones, and collateral circulation is made more difficult by the presence of the disease process in what might otherwise be the collateral blood vessels. When this fact is considered along with the frequency of anomalies in the same circulation, one can see that potential disaster is an ever present consequence of total occlusion.

In view of increasing interest during recent years in the surgical management of extracranial arterial stenosis or occlusion, it has become much more important to study such patients completely. Decisions regarding the accessibility and operability of lesions and prognosis of surgical intervention are much more easily achieved if proper angiographic opacification of all the vessels in this portion of the arterial tree is obtained in advance.

Arteriography as a technique has contributed a great deal of otherwise unobtainable information regarding the presence and adequacy of collateral circulation. This is especially true in regard to understanding two entirely different, but related, aspects of cerebral ischemia:

1. Variability of the clinical manifestations of cerebrovascular occlusion or stenosis,

2. Predictability of recovery from functional neurologic deficit.

From the technical point of view it is important to utilize, whenever possible, indirect injection methods rather than direct percutaneous puncture into diseased arteries. The indirect techniques can be accomplished either with or without catheter. When catheters are not used, it is advantageous to employ a pressure injector. One should avoid at all cost a direct puncture when there is risk of entering an area in which atherosclerosis is presumed to be present. The region in which the most grave risk is encountered is in the cervical carotid bifurcation. The patient with a short and broad neck particularly presents a problem. Under circumstances where this risk is ignored, there may be fragmentation of the plaque as a result of trauma from the needle and pieces of the plaque may be broken off and embolize down stream into the more distal cerebral vessels. On occasion this kind of debris from fragmented plaques in proximal vessels can produce an embolic stroke and an irreversible neurologic deficit.

In the past there has been considerable apprehension about employing arteriography on the part of many physicians which has caused them to avoid comprehensive studies, particularly in elderly persons with advanced atherosclerosis. This is no longer a tenable position since it has been shown that with refined techniques and new less irritant contrast media that the risks are minimal if the arteriographer is experienced and the necessary precautions are taken at the time of examination. During the procedure, one must avoid any maneuver which could produce lowering of the mean arterial blood pressure since this might contribute to the production of transient or even prolonged cerebral ischemia.

In summary, the following are suggested as guidelines for studying collateral circulation in patients with occlusive cerebrovascular disease.

1. Employ indirect injection techniques whenever feasible and avoid direct puncture of diseased arteries.

2. Use biplane serial angiography to reduce the amount of contrast and number of injections required for complete visualization.

3. Position the patient so as to be able to visualize the circle of Willis.

4. Obtain as much information as possible about direction and velocity of flow from one part of the circulation to another by selective catheterization and injection of each main trunk separately.

References

FIELDS, W. S., BRUETMAN, M. E., WEIBEL, J.: Collateral circulation of the brain. Monogr. surg. Sci. **2**, 183 (1965).
— EDWARDS, W. H., CRAWFORD, E. S.: Bilateral carotid artery thrombosis. Arch. Neurol. Psychiat. (Chic.) **4**, 369 (1961).
— RATINOV, G., CAMPOS, R. J.: Survival following basilar artery occlusion. Arch. Neurol. Psychiat. (Chic.) **15**, 463 (1966).
FISHER, C. M.: Occlusion of the internal carotid artery. Arch. Neurol. Psychiat. (Chic.) **65**, 346 (1951).
HAYREH, S. S., DASS, R.: The ophthalmic artery. I. Origin and intra-cranial and intra-canalicular course. Brit. J. Ophthal. **46**, 65 (1962).
PADGET, D. H.: The circle of Willis: Its embryology and anatomy. In: Intracranial arterial aneurysms, p. 67. (DANDY, W. E., Ed.). Ithaca, N. Y.: Comstock 1944.
QUIRING, D. P.: Collateral circulation. Anatomical aspects. Philadelphia: Lea & Febiger 1949.
RIGGS, H. E., RUPP, C.: Variation in form of circle of Willis. The relation of the variations to collateral circulation: anatomic analysis. Arch. Neurol. Psychiat. (Chic.) **8**, 24 (1963).
SEDZIMIR, C. B.: Angiographic test of collateral circulation through the anterior segment of the circle of Willis. J. Neurol. Neurosurg. Psychiat. **22**, 64 (1959).
WEIBEL, J.: Atlas of arteriography in occlusive cerebrovascular disease. Stuttgart: Thieme 1969.
— FIELDS, W. S.: Arteriography of the posterior cervicocranial circulation. Amer. J. Roentgenol. **98**, 660 (1966).

Some Basic Patterns of the Collateral Circulation of the Cerebral Arteries*

K. J. Zülch

It is the intent of this paper to illustrate that parts of the normal cerebrovascular circulation, when stenosed or occluded, may be taken over by others. The morphological anlage for the origin of such an auxiliary circulation and the resulting patterns will be described. The embryonic development is known from the textbooks and has been discussed by FIELDS (1965). Anomalies have to be distinguished from "secondary" channels.

If one vertebral artery is very small or absent, the caliber of the other may help to decide whether the larger vessel is the result of a discrepancy in the anlage (see ZÜLCH, 1969, Fig. 14) or at least partly the result of an increasing demand in the case of stenosing disease.

However, I shall refrain from giving any details here, but rather try to analyse the ground patterns of the resulting vascular channels.

Terminology: When we describe the main collateral pathways of the "four vessels" leading to the brain we get into difficulties with *terminology*. In clinical jargon, as in neuroradiology, we are accustomed to use the terms "collateral circulation" and "anastomoses" synonymously, yet they have actually different meanings. Anastomoses mean network-like *intercommunications of one or many supply systems*, where the direction of the pathway and the caliber are not defined. "Anastomosis" may be even used as a technical term. "Collaterals", on the other hand, are morphological pathways which run parallel and can substitute for each other (in case of deficiency of one!), in other words, *they are parallel supply systems to one organ*.

This is particularly true for the arterial channels around the joints—as the elbow joint—in case of impedance of the *medial* channels by flexion. Then the *lateral* pathways take over.

If we would apply such a strict definition to our subject, only the two vertebrals could suit this definition, both running parallel to the same organ and being able to substitute for each other when of normal calibre (Fig. 1).

Thinking in teleological terms, this is a wise arrangement, since the possibility of a mechanical strangulation is very great, both at the entrance into the foramen costo-transversarium and later in the vertebral column itself, as for example near the atlanto-occipital joint and while piercing the dura (see KRAYENBÜHL and YASARGIL, Fig. 149 a, b). This is particularly true when hyperostoses in cervical osteochondrosis are formed near the foramina costotransversaria (see ZÜLCH, 1970, Fig. 19).

In summarising: *Collaterals* are—by their original definition—double, triple or multiple parallel supply systems substituting for each other in case of deficiency;

* From the Max-Planck-Institut für Hirnforschung, Department of General Neurology, and the Neurological Department of the Städtische Krankenanstalt Cologne-Merheim.

anastomoses, on the other hand, are intercommunications of a network character in one or between two or more functionally separate systems, allowing the possibility of draining blood from them. An auxiliary supply may result, usually after widening of the channel, and flow may result in any direction.

Inspite of these logical distinctions in the use of the above two terms, both are introduced into our field as synonyms; but we will use them in the following discussion as close to their original definitions as possible.

The causes that bring of collateral channels or anastomoses into action. The causes are 1. a *temporary* or 2. a *permanent* impedance of an arterial channel. We will see that some of these *temporary* situations for the action of collaterals or anastomoses may be

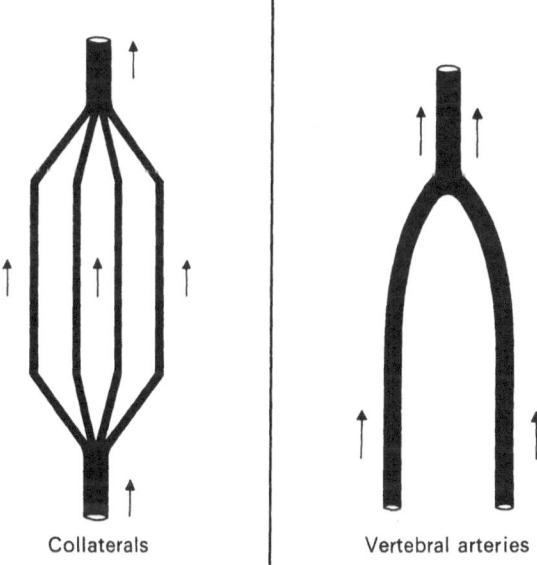

| Collaterals | Vertebral arteries |

Fig. 1. The only example of true "collaterals" is in the field of cerebral circulation: the two vertebral arteries

"physiological". This may be in the case of strangulation of one carotid in extreme rotation or extension of the neck, where the circle of Willis will compensate or by extreme positions at the atlanto-occipital joint where one vertebral artery supports or substitutes for the supply of the other.

Moreover, some congenital or aquired variations of the "cerebral" arteries may temporarily impede flow in extreme positions and convert into functional "stenoses", as is caused by *tortuousity*, *coiling* or *kinking* of the great arteries.

According to FIELDS tortuousity refers to a congenital lesion; coiling may be congenital or acquired and kinking is always a sign of arterial disease (arteriosclerosis), which, however, may also increase the congenital variations of tortuousity and coiling. Their effect on flow may be shown in the various positions of head and neck during angiography. On the other hand, a *permanent* change of the arterial circulation is usually due to disease: arteriosclerosis, thrombosis or embolism. Arteriosclerosis may occur in the cerebral vessels in a stenosing and/or ectatic form, which may then lead to final occlusion by increase of the arteriosclerotic plaque or by a terminal thrombosis in such a stenosed vessel.

The resulting anastomotic systems will be better developed if the stenosis or occlusion begin *slowly* and will depend on the hemodynamic situation. This will be discussed later.

Anastomoses and collaterals. Two ways exist to describe the substitute arterial communications important for the neuroradiologist and clinician: a) the separate examination of each of the "four vessels" from its point of origin to its termination. Here a case description would be appropriate. b) The second way would be a systematic analysis of the basic pattern of such channels used for an auxiliary circulation. This will be used in the following discussion.

The Basic Patterns: I. Temporary Use of a Parallel Pathway

The posterior inferior cerebellar ↔ superior cerebellar anastomosis. The simplest pattern is the temporary use of a *parallel running arterial* pathway, in order to circumvent a

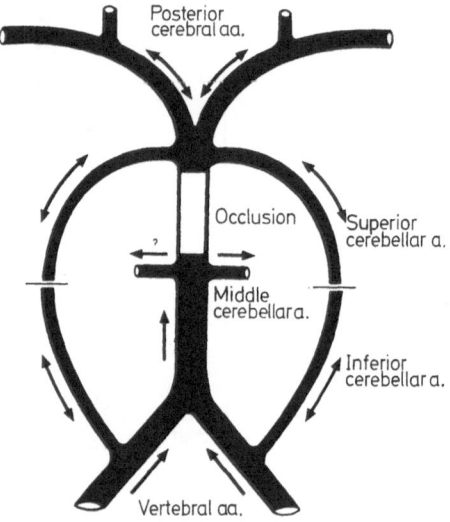

Fig. 2. In an occlusion of the basilar artery anastomoses between the superior and inferior posterior cerebellar arteries begin to function

block, if this is a true "collateral" like the vertebral artery. Naturally there will be no reversal of flow. This is different if an anastomosis is used. The best example for this pattern is the anastomotic system between the superior cerebellar and posterior inferior cerebellar arteries in case of blockage of the middle segment of the basilar artery (Fig. 2).

This may even be sufficient to prevent the "syndrome of basilar thrombosis" with the catastrophe of tetraplegia etc.

The external carotid → ophthalmic → internal carotid anastomosis. More complex is a similar pattern to bypass a blocked internal carotid artery via the ophthalmic anastomoses of the external to the internal carotid artery.

The ophthalmic artery supply territory branching from the internal carotid normally anastomoses with the external carotid artery by 1. the frontal artery via the superficial temporal, 2. the ethmoidal via the internal maxillary, 3. dorsal nasal artery via the external maxillary. Even the middle meningeal may participate (Fig. 3, 4).

A retrograde flow through the ophthalmic artery fills the internal carotid and its terminal arteries. Moreover and to our surprise, it also usually fills retrogradely the syphon of the carotid down to the entrance through the base of the skull.

According to Fields, this is due to the existence of the carotico-tympanic artery, which according to him, fills the carotid stump from the external (see his Fig. 37 a, 1965). In our cases, we have never been able to see such an artery even in subtraction (Fig. 5). I have (1970) a different explanation: that the carotid is filled retrogradely by the ophthalmic down to where the arteries to the pituitary gland and trigeminal ganglion (see Pernkopf, Fig. 66 b) are given off.

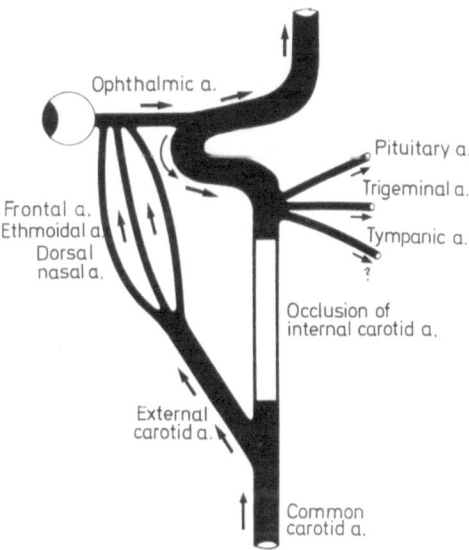

Fig. 3. In case of occlusion of the internal carotid artery, anastomoses between the external and distal internal carotid artery form a detour around the occlusion

Occasionally the hypertrophied pulsating superficial temporal artery of this ophthalmic collateral circulation may be seen with the naked eye and its increased pulsation noticed by the patient (Zülch, 1963, Fig. 4).

The vertebralis → anterior spinal → vertebralis anastomosis. The same ground pattern of a temporary "parallel" detour which finds access again to the original supply system is seen when both vertebral arteries are blocked distally near the entrance into the cranial cavity. Then the segmental branches from the vertebral artery to the anterior spinal are used to supply this artery (Fig. 6), which communicates with both vertebrals intracranially (Schechter, 1964; Fields, 1965, Fig. 35). All of these arteries dilate very markedly; a conversion of flow is only seen in the upper part of the anterior spinal artery, where, however, already normally the direction of flow changes according to the body position (upright, lying—so called "Teilströmchen-theorie").

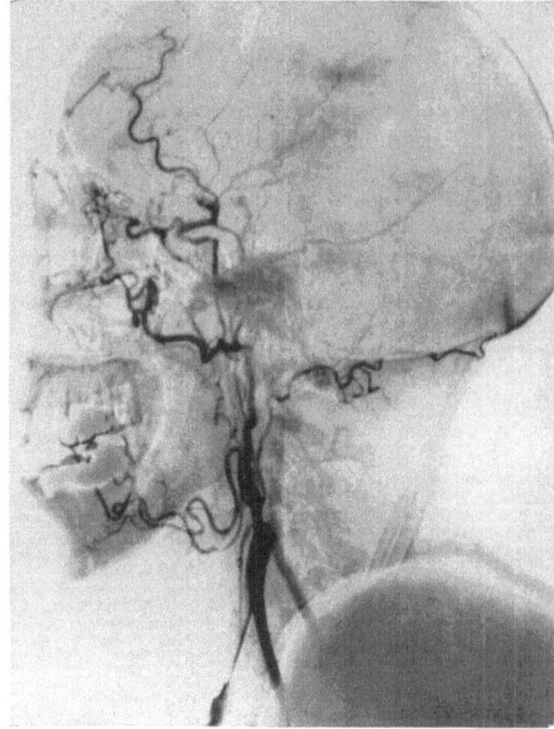

Fig. 4

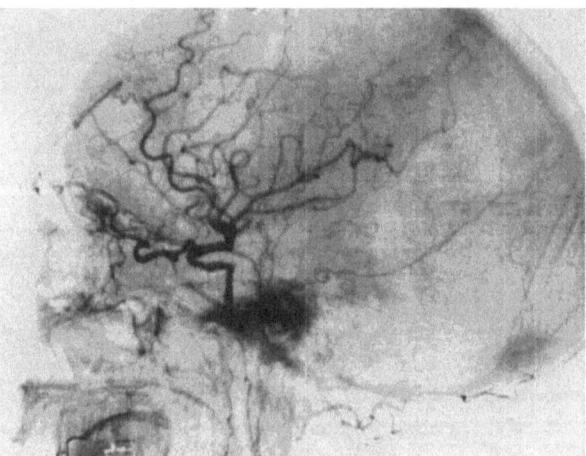

Fig. 5

Fig. 4 and 5. Angiogram in case of a right sided occlusion of the internal carotid with a supplementary circulation via the ophthalmic artery according to Fig. 3. — Fig. 5 shows a reflux into the occluded stump of the carotid artery near its entrance into the base of the skull

II. The Transverse Anastomoses between Two Equal Supply Systems

Here we have again to distinguish between temporary "physiological" and permanent "pathological" supply patterns. The most common "physiological" transverse system is the circle of Willis. This system seems predisposed for the purpose of distribution (Fig. 7) between three different and equal systems in case of temporary deficiency of

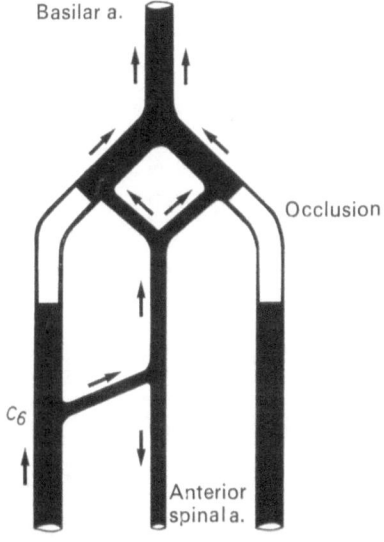

Fig. 6. In an occlusion of both vertebrals, the spinal anterior artery is used as a detour to the blockage

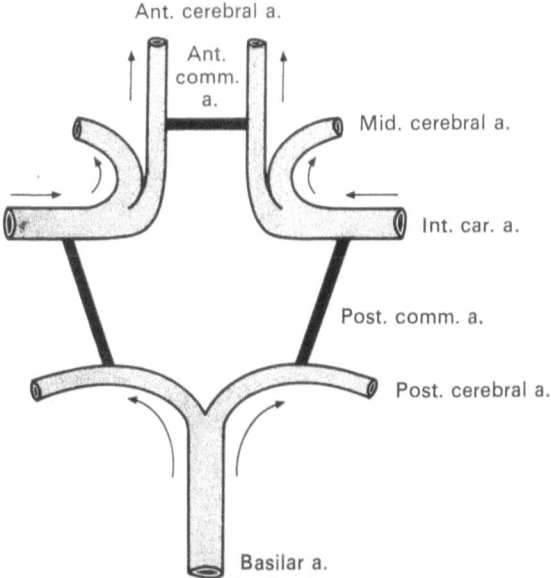

Fig. 7. The three independent systems of the two carotids and the vertebro-basilar system are anastomosed by the circle of Willis

one of them: the two carotids and the vertebro-basilar artery. How often and how far this is needed already under "physiological" conditions (extreme turning or bending of neck and head with strangulation of a carotid or a vertebral artery) is not sufficiently investigated yet. In some of our angiograms different parts of this annular

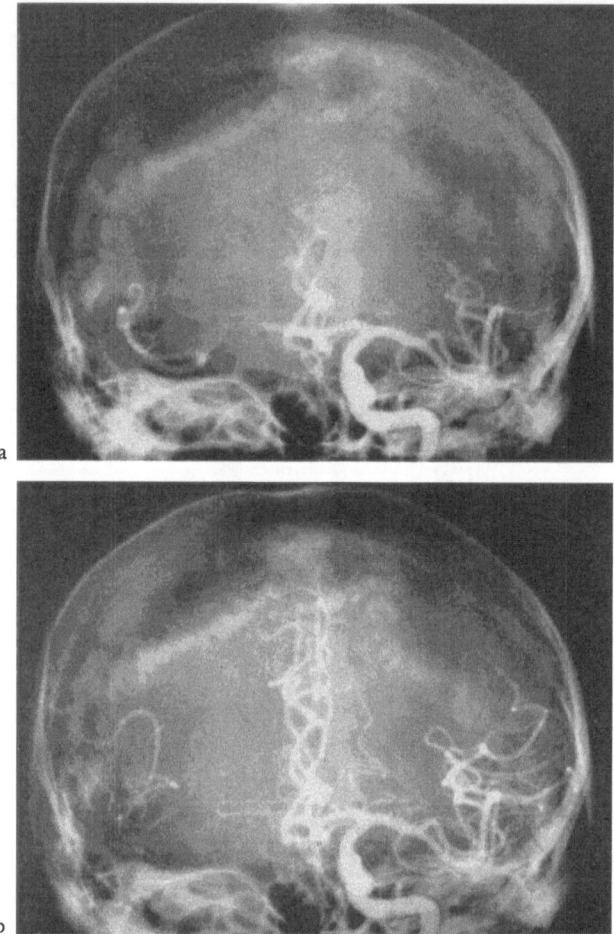

Fig. 8. Occasionally a bolus of the contrast medium is injected to the contralateral side through the anterior communicating artery (in diastole?)

system may be made visible; however, without much understanding of the hemo-dynamic conditions responsible.

Early neurosurgical investigations postulated that the pressure in the system of Willis up to the terminal great cerebral arteries may not be increased by the injection pressure during carotid or vertebral angiography. However, TÖNNIS and SCHIEFER already reported that occasionally a small bolus of contrast medium may be temporarily injected from the vertebrobasilar system into the carotid via the posterior communicating arteries (their Fig. 73).

We have several times noticed a bolus passing from one carotid system to the other (through the anterior communicating artery to the other middle cerebral). This probably occurred during diastole where the pressure gradient was changed by the injection pressure, inspite of the aforementioned data (Fig. 8 a, b).

Unfortunately the circle of Willis is normal only in 25% (Riggs 19%) whereas in another 25% it may be "open" and its function as a redistributing system deficient; in the remaining cases, blocking of one of the three great supply systems may have unexpected and dramatic consequences due to hypoplasia of parts of the circle of Willis [see Fig. 51 of Krayenbühl and Yasargil (1957) after Padget or Fig. 29, Fields, 1965 etc.].

Meanwhile, "tests" for the anastomosing function of the circle of Willis have been described for angiographic (Krayenbühl and Yasargil, p. 47; Decker) or electro-encephalographic (Gastaut and Behrend, 1961) diagnosis.

It will be impossible to show here all the different possibilities of intercommunication by the circle of Willis seen angiographically after carotid or brachial retrograde angiography. Sometimes these passages follow rules which cannot yet be explained.

Loeb and Favale described cases of arteriovenous aneurysms, where an electrobiological "focus" was seen in the *contralateral hemisphere*. Fazio and Loeb called this phenomenon a "steal" and, moreover, the attitude of the "donor" hemisphere "altruistic". Apparently, the supplementary share to the malformed circulation was too great and detrimental to the normal side. Hemodynamically this may be understood because of the immense widening of all the channels supplying the a.v.m., which lacked a capillary network, so that the flow was several times greater than normal.

We have observed a different condition which, however, led to the same unfavorable consequences, namely a cerebrovascular insufficiency in the healthy "normal" hemisphere, whereas the hemisphere with occlusion of its vascular supply did not suffer.

In a patient aged 64 a focal seizure of the right arm occurred and an electrobiological focus was found in the left hemisphere. Angiography showed filling of both middle and anterior cerebral arteries from the left carotid. This was interpreted as a probable consequence of a right carotid artery occlusion and right brachial angiography was performed. An occlusion of the right common carotid was visualized.

The hemodynamics of this process were not understood. Even if one presumed a maximum dilatation of the vessels on the right side, the resulting insufficient vascular region on the left ought to have led to the same dilatation, so that, according to the phenomenon of the "last meadow", the focus would be expected in the most distal part of this circulation, i.e. the posterior temporal region of the right side. Even the "posterior" circulation did not give the "clue" to this phenomenon. That in occlusion of one carotid the sequelae may sometimes be distributed equally to both hemispheres was already suggested by Zülch and Herberg (1949, Fig. 1), where a pneumogram showed bilateral equal hydrocephalus after unilateral carotid artery ligation. —

The next typical example of "transverse" anastomoses between two different and divergent systems is one of the most complicated detours:

The subclavian → vertebral → vertebral → subclavian anastomosis (the so called "subclavian steal" phenomenon.)

Because of the "steal" phenomena occasionally noticed (Contorni, Reivich, Fields, Vollmar etc.), such cases received the clinical jargon-term "subclavian steal". This arterial detour has been very popular since the discussion of this "steal" as a possible cause of vertebro-basilar insufficiency.

FIELDS, however, pointed out at the 4th Salzburg Conference that in the majority of cases there are no such "steal" phenomena; consequently the name is incorrect and likewise there is no indication for surgery. Only in some cases (6 out of 7 reported by NORTH *et al.*, 50% in other reports of the literature) simultaneous vertebro-basilar insufficiency was seen in the course of muscular action of the deficiently supplied arm.

In one of our cases of such a subclavian detour-circulation via the two vertebrals there was also no "steal" since the arm with subclavian insufficiency was paralysed by a cerebral infarct after a simultaneous carotid blockage (see Fig. 3 to 5). This detour

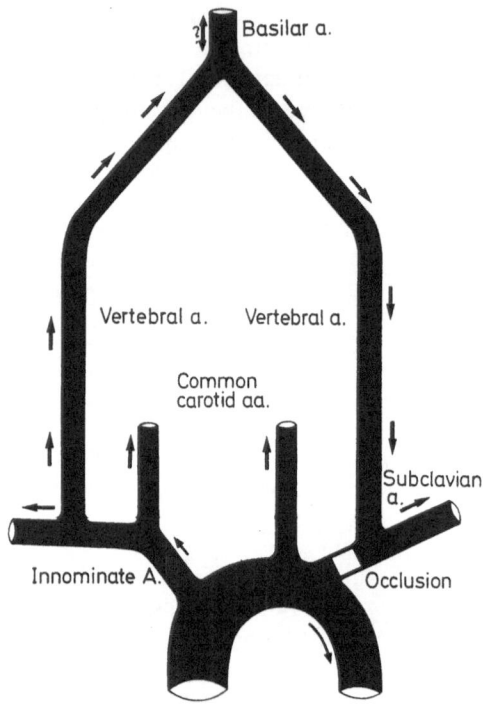

Fig. 9. The confluence of the two vertebrals may give rise to a very long anastomotic supply circulation in case of occlusion of one subclavian artery. This may even be detrimental to the vertebro-basilar circulation

is by far the longest "collateral circulation" in the field of the "four cerebral vessels", its entire length being at least 40 cm.

The transverse anastomoses between the two external carotids. The two external carotid systems are interconnected by a particularly dense network of anastomoses in their terminal supply areas. If only the common carotid is blocked, the external remains open as, for instance, in cases of surgical ligature, there will immediately be inter-communication between the external carotids and a retrograde flow into the internal carotid, the blood stemming from the transverse anastomoses (Fig. 10) between the two superior thyroid arteries, lingual, ascending pharyngeal, internal maxillary, mental, coronal, labial, inferior angular arteries etc. (see FIELDS, 1965, Fig. 9 and Fig. 2 to 10, as well as tables page 230/232).

These transverse anastomoses between the two external carotids have been shown to be dilated by NIETEN (1893) in morphological preparations of such patients at autopsy.

The transverse anastomoses between the carotid and vertebral systems: The external carotid ↔ occipital ↔ vertebral anastomoses. Other transverse anastomoses exist extra-

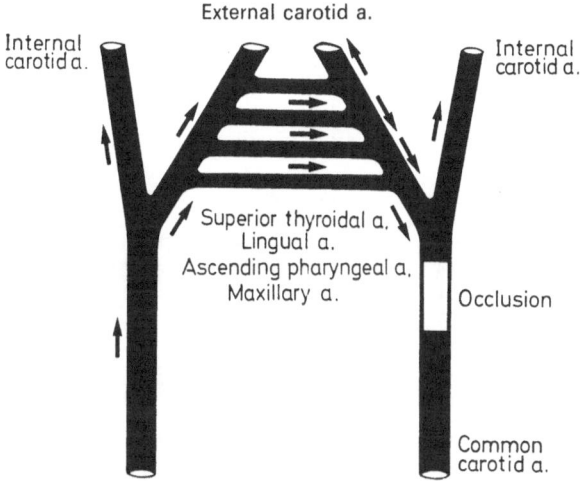

Fig. 10. The two parallel systems of the external carotid arteries are interconnected by numerous transverse anastomoses

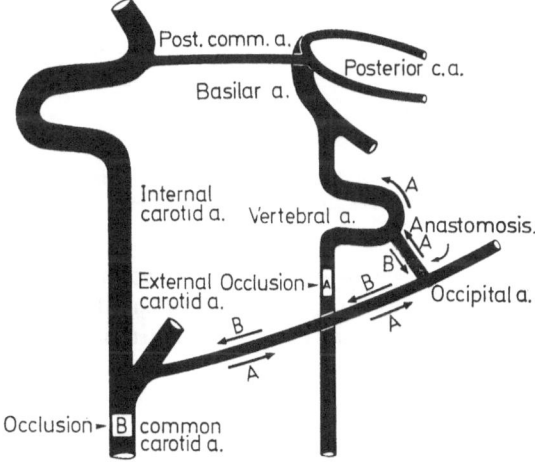

Fig. 11. The vertebral and the carotid systems are anastomosed by a greater arterial channel, which can work either way (A. or B.) in occlusions

cranially between the carotid and the vertebro-basilar system. Here the undamaged systems are—as always—drained by the deficient, whereby intercommunication is possible either way. In this case, it is achieved by an artery anastomosing between the occipital and vertebral arteries which is very regularly present and is well described by the anatomists (SOBOTTA; CORNING; HAFFERL; RAUBER-KOPSCH; TÖNDURY, 1951; SCHULZE and SAUERBREY, 1956); when the vertebral was occluded it was shown in angiography of the external carotid by Schürmann (1954). We have similar pictures in a case of occlusion of the vertebral system (Fig. 11) of one side, where the external carotid system is the supply source. But also the reverse flow exists, where in occlusion of the common carotid we find a draining of the vertebral artery via the occipital anastomosis. Thus the internal carotid can be filled even beyond the syphon.

Apart from this direct arterial intercommunication between the external carotid and the vertebral arteries, terminal muscular branches of the external carotid (Krayenbühl and Richter, Fig. 100) also exist. Or the rather parallel thyrocervical plexus may anastomose with the vertebral artery. It may be used to supply the vertebro-basilar system, as shown by a recent personal observation, where the vertebral was occluded at the tuberculum anterius, but refilled distal to this point by the afore mentioned anastomoses.

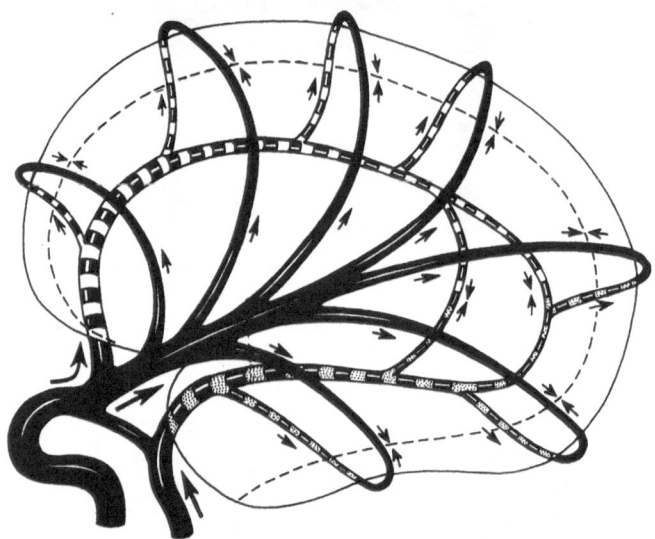

Fig. 12. A network-like arterial anastomotic system is provided by the meningeal anastomoses of Heubner between the three great cerebral arteries

Transverse intercommunications between the great intracranial arteries: Heubner's *"meningeal anastomoses"*. We owe to the elder Heubner (1874) the observation that practically all the great arterial cerebral and cerebellar arteries form a network (Fig. 12) on the surface of the brain, the anastomotic channels being smaller arteries sufficient in size to supply blood to the deficient regions in case of emergency. Actually, the first description may be found with Ruysch (1699).

Cohnheim's theory, that the brain arteries were "end arteries", has to be rejected since then and is only valid for the arteries of the basal ganglia and the brain stem (Heubner's anterior circumflex artery, the lenticulo-striate and lenticular-optic as well as the anterior chorioidal artery, moreover the tegmental, pontine and oblongata branches). Only the anterior chorioidal may have a certain greater intercommunication with its "posterior" partner (i.e. post. chorioidal artery) which, however, may vary in size and function. The perforating branches, and particularly, the median, paramedian, short and long circumflex arteries of the brain stem also belong to this category, having practically no anastomoses after entering the brain.

Heubner's concept of a reticular arterial network on the surface of the brain was, however, long rejected (Charcot) although confirmed by anatomists (Testut, Beevor).

Even though French clinicians in the beginning of this century reconfirmed HEUBNER's findings and showed the decrease of the size of an infarct due to the action of "HEUBNER's meningeal anastomoses" (LECENE and LHERMITTE, J.). Moreover, TEMPLE FAY proved the existence of these arterial intercommunications by his own work, but it was not until 1951, that they were "rediscovered" by VAN DER EECKEN and ADAMS. The first angiographical proof of such anastomotic channels is due to FISCHER-BRÜGGE's angiogram of 1944, published in 1949 (his Fig. 1, 2) and honoured in the meantime by giving it his name as the "corpus callosum anastomosis" of FISCHER-BRÜGGE (see below).

The various types of possible intercommunications by the "meningeal anastomoses" are excellently described by VAN DER EECKEN and ADAMS and by VAN DER EECKEN (1959). If we try to analyse the basic pattern acting in most of the cases of either middle cerebral or anterior cerebral occlusions, it can be understood as the "draining" of a *parallel but separated neighbour system* by "transverse anastomoses" where in the occluded artery a complete reversal of flow results (Fig. 13). For instance,

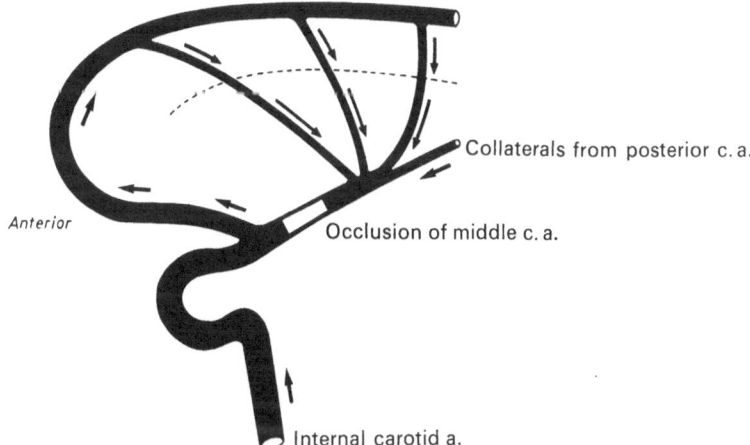

Fig. 13. In occlusion of the middle cerebral artery the parallel systems of the anterior and posterior cerebral arteries are drained by "HEUBNER's meningeal anastomoses"

in a block of the middle cerebral the blood may be drained from the anterior as well as from the posterior cerebrals, via HEUBNER's meningeal anastomoses. In case of the occlusion of the anterior cerebral arteries, the posterior system is particularly used for drainage by the corpus callosum anastomoses which work in either direction between the vertebro-basilar and the carotid systems.

Another form of very common "transverse" anastomoses between two different, though parallel systems, exists between the two anterior cerebral arteries on the top of the corpus callosum (VAN DER EECKEN, Fig. 3 A; ZÜLCH, 1964, Fig. 4). They are well known to neuroradiology and belong to the system of "meningeal anastomoses".

III. The Micronetwork of Ring Anastomoses

The arachnoidal arterial ring systems. These little arterial rings on the surface of the brain were well studied by SCHMIDT, yet they do not seem to take up any considerable supply service in case of emergency. They do not fit into any auxiliary patterns described up to now. However, in *chronic* obliteration of the lumens of smaller surface arteries (250 to 750 μ lumen)—as in the case of thrombangiitis obliterans v. WINIWARTER-BUERGER—they form an anastomotic network (Fig. 14), which macroscopically resembles "hyperemia" of the meninges but is, however, of "arterial" origin and different from the usual venous "hyperemia" (ZÜLCH, 1966, Fig. 3, 1969, Fig. 4).

Cerebral thrombangiitis obliterans (or v. WINIWARTER-BUERGER's disease) is certainly ill defined up to date and clinically diagnosed far too commonly. Morphologically, we had two cases in about 500 brains with vascular disease. On the base of the morphological findings—consisting of thromboses, fresh and organised, in the arteries of $^1/_4$ to $^3/_4$ mm diameter and which are pseudosystematically distributed over the brain and cerebellum. I have risked (in 1966 at the International Academy of Angiology in Madrid) to draw a hypothetical angiogram of the cerebral form of v. WINIWARTER-BUERGER's disease. We have, however, never seen such a case angiographically, which was apparently first demonstrated by HACKER (1968). His angiogram shows the breaking up of the little terminal arteries in the middle

Branches of anterior and posterior c. a.

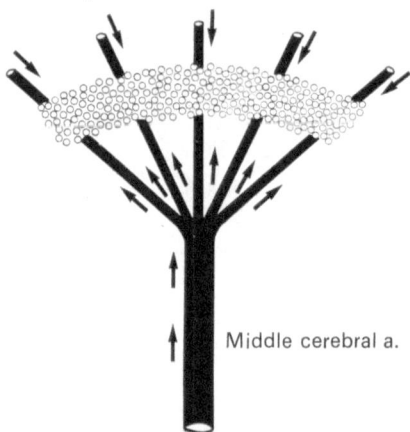

Middle cerebral a.

Fig. 14. The "arachnoidal arterial ring" anastomoses are used as a substitutional supply in case of distal blockage of the great branches of the cerebral arteries in thrombangiitis obliterans

cerebral territory on one side and the anterior on the other. In the border-line zone a "blush" is visualised demonstrating the hyperemia of the arachnoidal rings. By this demonstration the discussion of the angiogram of v. WINIWARTER-BUERGER's disease may reach its final stage (see ZÜLCH, 1969).

IV. The "Capillary Anastomoses"

These *transverse intracerebral* microchannels so well studied by R. A. PFEIFFER ("Pfeiffer's capillary anastomoses") do not have any significant function in the supply of blood to a deficient area in case of *emergency*. This is very detrimental for the "brainstem" where the arteries seem to have no major intercommunications apart from the capillaries. In case of an occlusion of a lenticulostriate or a paramedian pontine artery at its origin, a total infarct of the supply area ought therefore to result (see above). The explanation, why the morphological findings do not correspond to such a concept has not been satisfactorily explained. It may be that in chronic stenoses the capillary anastomoses begin to work and decrease the size of a possible infarct. This brings to an end the description of the morphological ground patterns of the collaterals and anastomoses of the "four cerebral basic vessels".

The functional prerequisites for collateral and anastomotic auxiliary circulation. In this context two questions remain to be answered, equally important to the clinician and to the neuroradiologist: a) When do these systems of intercommunication develop and b) to what degree may they be enlarged? Four predispositions seem to be necessary.

1. The sufficient *morphological anlage* of collaterals and anastomoses. An "open" circle of Willis cannot fulfill its function even if the hemodynamic conditions are optimal.

2. The *time development* in the case of emergency. Since most of the anastomoses form "physiological stenoses" (ZÜLCH, 1967, Fig. 3) between the two supply systems, the lumen must be sufficient to allow of a transport action in shortest time in a sufficient quantity. Therefore, a slowly growing arteriosclerotic stenosis favours the "widening" of any anastomotic circulation, whereas the rapid occlusion of an artery

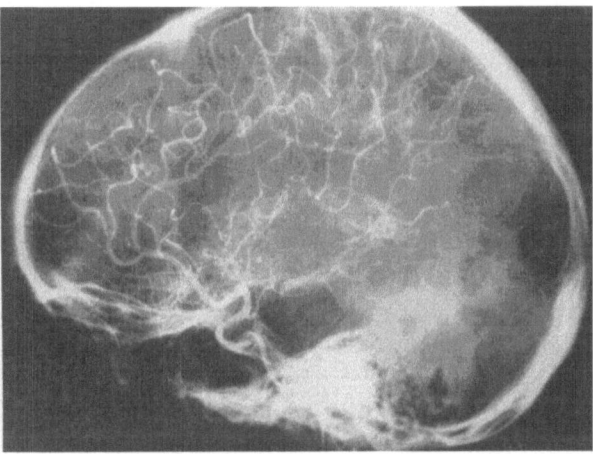

Fig. 15. The system of the Sylvian arteries occluded by proximal thrombosis is filled by reflux from the HEUBNER's meningeal anastomoses. Reactive hyperemia of the region and "early" deep veins (see text)

by embolism may only count on a preexisting and sufficiently large anastomotic system like the circle of Willis. Thrombosis on the other hand, may stand in between these two time developments since often 24 h or more are needed for its formation, or else it is superimposed on a stenosing arteriosclerotic plaque, which has already introduced a collateral supply before an emergency situation.

3. The *hemodynamics* must be sufficient to overcome any "narrow" parts of the anastomotic communications, i.e. the arterio-arterial or even arteriolar-arteriolar anastomoses; the blood pressure—as has been shown convincingly—must be high enough to let the blood pass these and any other physiologically or pathologically stenosed parts. This is known since the observations (ZÜLCH and HERBERG, 1949) of war lesions of the carotid associated with great loss of blood and shock and other traumatic vascular lesions.

4. *Arteriosclerosis* must *not* have changed the arterial wall to such a degree as not to allow any "widening" (which is seen gradually developing in every arteriovenous shunt).

Does the existence of certain collateral and anastomotic systems give information about prognosis? This question has been discussed in the literature of the ophthalmic collateral circulation. Two observations of our own may throw some light on this problem and show how difficult it is to gain such information neuroradiologically. In the case of a 49 years old doctor, thrombosis of the middle cerebral artery probably developed during the clinical observation of two days. Angiography on the 4th day showed excellent meningeal anastomoses from the anterior cerebral artery refilling the Sylvian system up to the occluded stump. Yet, the hemiplegia was not later reversible. On the other hand, in the case of a lady of 59 years with a similar angiogram and equal meningeal anastomoses, only a very slight motor aphasia remained on testing although there was no difference in the angiogram of the two cases.

Summary: The principles of anastomotic and collateral pathways are discussed and the predispositions for their functioning emphasized.

References

BEEVOR, CH. E.: The cerebral arterial supply. Brain **30**, 403—425 (1907).

CHARCOT, J. M.: Leçons sur les maladies du système nerveux faites à la Salpêtrière. Paris: Delahaye 1872—1873

COHNHEIM, J.: Untersuchungen über die embolischen Prozesse. Berlin: Hirschwald 1872.

CORNING, H. K.: Lehrbuch der topographischen Anatomie. München: Bergmann 1949.

CONTORNI, L.: Il circolo collaterale vertebro-vertebrale nella obliterazione dell' arteria subclavia alla sue origine. Minerva chir. **15**, 268 (1960).

DECKER, K.: Der Schlaganfall als neuroradiologisches Problem. Dtsch. med. Wschr. **83**, 205 (1958).

DENNY-BROWN, D.: The treatment of recurrent cerebrovascular symptoms and the question of "vasospasm". Med. Clin. N. Amer. **35**, 1457—1474 (1951).

— Basilar artery syndromes. Bull. New Engl. med. Cent. **15**, 53—60 (1953).

— MEYER, J. S.: The cerebral collateral circulation. 2. Production of cerebral infarction by ischemic anoxia and its reversibility in early stages. Neurology (Minneap.) **7**, 567—579 (1957).

FAY, T.: The cerebral vasculature. J. Amer. med. Ass. **89**, 1727—1730 (1925).

FAZIO, C., FIESCHI, C., AGNOLI, A.: Discussion sur la pathogenese des infarctus cérébraux: Valorisation des facteurs hemodynamiques. Anales del XII Congreso Latinoamericano de Neurocirugia, Symposium Internacional de Investigaciones Neurologicas, pp. 267 to 280, Lima 1967.

FIELDS, W. S., BRUETMAN, M. E., WEIBEL, J.: Collateral circulation of the brain. Monogr. surg. Sci. **2**, 183—259 (1965).

FISCHER-BRÜGGE, E.: Der persistierende Hirnprolaps nach Schußverletzungen. Zbl. Neurochir. **9**, 18—45 (1949).

GASTAUT, H., BEHREND, R. CH.: Die klinische elektroencephalographische Differentialdiagnostik der fokalen und generalisierten epileptischen und ischämischen Anfälle. Acta neuroveg. (Wien) **23**, 137—153 (1961).

HAFFERL, A.: Lehrbuch der topographischen Anatomie. Berlin-Göttingen-Heidelberg: Springer 1953.

HEUBNER, O.: Die luetische Erkrankung der Hirnarterien. Leipzig 1874.

KRAYENBÜHL, H., RICHTER, H. R.: Die zerebrale Angiographie. Stuttgart: Thieme 1952.

— YASARGIL, M. G.: Die vaskulären Erkrankungen im Gebiet der A. vertebralis und A. basilaris. Stuttgart: Thieme 1957.

— — Die zerebrale Angiographie. Stuttgart: Thieme 1965.

LECENE, B., LHERMITTE, J.: Une observation anatomo-clinique d'un cas de ramollissement cérébral consécutif à l'obliteration de l'artère sylvienne gauche par embolie métallique. Rev. neurol. **27**, 1116—1121 (1920).

LHERMITTE, FR.: Les circulations de suppléance du cerveau. Leur rôle dans l'artériosclérose. Brux.-méd. **40**, 1267—1277 (1959).

— GAUTIER, J. C., DEROUESNE, C.: Anatomopathologie et physiopathologie des sténoses carotidiennes. Rev. neurol. **115**, 641—672 (1966).

·LOEB, E., FAVALE, F.: Contralateral EEG abnormalities in intracranial arteriovenous aneurysms. Arch. Neurol. Psychiat. (Chic.) **7**, 121—128 (1962).

MEYER, J. S., DENNY-BROWN, D.: The cerebral collateral circulation. 1. Factors influencing collateral blood flow. Neurology (Minneap.) **7**, 447—458 (1957).

— FANG, H. C., DENNY-BROWN, D. S.: Polarographic study of cerebral collateral circulation. Arch. Neurol. Psychiat. (Chic.) **72**, 296—312 (1954).

NIETEN: Dissertation, Bonn 1893.

NORTH, R. R., FIELDS, W. S., DE BAKEY, M., CRAWFORD, E. ST.: Brachial-basilar insufficiency syndrome. Neurology (Minneap.) **12**, 810—820 (1962).

PADGET, D. H.: The circle of Willis: Its embryology and anatomy. In: Intracranial arterial aneurysms, p. 67 (DANDY, W. E., Ed.). Ithaca, N. Y.: Comstock 1944.

PERNKOPF, E.: Topographische Anatomie des Menschen, Bd. IV/1: Der Kopf. München-Berlin-Wien: Urban & Schwarzenberg 1957.

PFEIFER, R. A.: Anastomosen der Hirngefäße. Dargestellt am asphyktisch-hyperämischen Kindergehirn. J. Psychol. Neurol. (Lpz.) **42**, 1—173 (1931).

RAUBER-KOPSCH: Lehrbuch der Anatomie des Menschen. Leipzig: Thieme 1948.

REIVICH, M., HOLLING, H. E., ROBERTS, B., TOOLE, J. F.: Reserval of blood flow through the vertebral artery and its effect on cerebral circulation. New Engl. J. Med. **265**, 878—885 (1961).

RIGGS, H. E., RUPP, CH.: Variation in form of circle of Willis. Arch. Neurol. Psychiat. (Chic.) **8**, 8—14 (1963).

RUYSCH, F.: Epistola anatomica, problematica, duodecima, authore Mich. Ernesto Ettmullero, etc., ad virum clarissimum Fredericum Ruysch, etc., de cerebri corticali substantia. Amsterdam 1699.

SCHECHTER, M. M., ZINGESSER, L. H.: The spinal arteries. 7th Symposium Neuroradiologicum New York 1964. Acta radiol. Diagn. **5**, 1124—1131 (1966).

SCHÜRMANN, K.: Darstellung der A. vertebralis und ihrer Äste im Angiogramm von der A. carotis externa aus. Zbl. Neurochir. **14**, 362—365 (1954).

SCHULZE, H. A. F., SAUERBREY, A.: Zur Frage der Anastomosen zwischen der A. vertebralis und der A. occipitalis. Zbl. Neurochir. **16**, 76—80 (1956).

SOBOTTA, J.: Atlas der deskriptiven Anatomie des Menschen, Band 3. Berlin-München-Wien 1946.

TESTUT, L., LATARJET, A.: Traité d'anatomie humaine, 8e. Ed. Paris: Doin 1929.

TÖNDURY, G.: Angewandte und topographische Anatomie. Stuttgart: Thieme 1951, 2. erweiterte Auflage 1959.

TÖNNIS, W., SCHIEFER, W.: Zirkulationsstörungen des Gehirns im Serienangiogramm. Berlin-Göttingen-Heidelberg: Springer 1959.

VAN DER ECKEN, H. M.: The anastomoses between the leptomeningeal arteries of the brain. Springfield: Ch. C. Thomas 1959.

— ADAMS, R. D.: The anatomy and functional significance of the meningeal arterial anastomoses of the human brain. J. Neuropath. exp. Neurol. **12**, 132—157 (1953).

VOLLMAR, J., EL BAYAR, M., KOLMAR, D., PFLEIDERER, TH., DIEZEL, P. B.: Zerebrale Durchblutungsinsuffizienz bei Verschlußprozessen der A. subclavia (subclavian steal effect). Dtsch. med. Wschr. **90**, 8—14 (1965).

WEIBEL, J., FIELDS, W. S.: Atlas of arteriography in occlusive cerebrovascular disease. Stuttgart: Thieme 1969.

ZÜLCH, K. J.: Gedanken zur Entstehung und Behandlung der Schlaganfälle. Dtsch. med. Wschr. **89**, 1524—1530, 1585—1590 (1960).

— Die Pathogenese von Massenblutung und Erweichung unter besonderer Berücksichtigung klinischer Gesichtspunkte. Acta neurochir. (Wien), Suppl. **VII**, 51—117 (1961).

— Zur Pathogenese des cerebrovaskulären Insultes. Internist **4**, 64—70 (1963).

— Morphology and pathogenesis of cerebral infarction. Anales del XII Congreso Latinoamericano de Neurocirugia, Symp. Internacional de Investigaciones Neurologicas, p. 255—265. Lima 1967.

Zülch, K. J.: Allgemeine Prinzipien bei der Entstehung der Kollateralkreisläufe der Hirn-
arterien. Radiologe 9, 396—406 (1969).
— The cerebral form of v. Winiwarter-Buerger's disease. Does it exist? Angiology 20,
61—69 (1969).
— Reconsiderations of the clinical problem of cerebrovascular insufficiency. In: Research
on the cerebral circulation. III. Intern. Salzburg Conference, p. 1—41. (Meyer, J. S.,
Lechner, H., Eichhorn, O., Eds.). Springfield (Ill.): Charles C. Thomas 1969.
— Angiographische Befunde zur Pathogenese der Hirndurchblutungsstörungen. Zbl.
Neurochir. 31, 1—25 (1970).
— Herberg, H.-J.: Das klinische Bild der akuten Blutsperre der Arteria Carotis. Dtsch. Z.
Nervenheilk. 160, 38—70 (1949).
— Kleihues, P., Gabe, D.: Die aktuelle Problematik auf dem Gebiet der Pathogenese,
Klinik und Therapie der Hirndurchblutungsstörungen. II. Int. Salzburger Konferenz,
S. 339—367. Wien: Verlag Brüder Hollinek 1964. — Wien. med. Wschr. 116, 494—503
(1966).

Functional Studies
on Normal and Pathological Cerebral Blood Flow

Fundamental Aspects of the Human Cerebral Circulation in Cerebrovascular Disease*

S. KETY

Measurement of cerebral blood flow in man by the exchange of an inert gas between blood and brain has been carried out over the past twenty years by a number of groups utilizing various adaptations of the underlying principles [1—10]. Modifications have been introduced which have increased the convenience or accuracy of the measurements or have yielded continuous information. It may be of interest to review some of the knowledge of the cerebral circulation which has thus accumulated, especially those aspects which have made for a better understanding of the pathophysiology, pathogenesis, diagnosis and treatment of cerebrovascular disorders.

Fundamental Physiology

Much of the new information relating to the human cerebral circulation has been at the basic physiological level—in fact, it has been said that more basic knowledge in this area has been acquired through studies in man in the past twenty years than was ever gained in lower animals. Although all fundamental information inevitably enriches the scientific background of the clinician, certain concepts are of special relevance to cerebrovascular disease.

In opposition to the older view, largely promulgated by LEONARD HILL [11] but persisting for more than a generation, that the cerebral circulation passively followed the blood pressure, the modern concept recognizes that a certain minimum level of mean arterial pressure (about 80 mmHg) is necessary for the maintenance of normal cerebral perfusion. This is provided by the normal relationships between cardiac output and peripheral resistance and zealously guarded by the circulatory reflexes, but as long as the pressure does not fall below this level, the rate of blood flow to the various parts of the brain is actually intrinsically controlled [12, 13].

At least two possibilities exist for the intrinsic regulation of cerebral circulation—neurogenic and humoral. Although there is a well-defined autonomic innervation of cerebral vessels arising from the carotid and vertebral plexuses which are in turn provided with sympathetic and parasympathetic supplies [14], the role of these neuro-

---

* From the Harvard Medical School Psychiatric Research Laboratories, Massachusetts General Hospital.

genic mechanisms remains obscure [15]. Studies on the effects of autonomic stimulation in animals [16] or of cervical sympathetic blockade or ganglionectomy in man
are equivocal [17, 18]. On the other hand, certain products of cellular metabolism,
notably carbon dioxide, have powerful dilator effects on cerebral arterioles [19] and
seem capable of accounting for most of the autoregulation within the cerebral circulation. Thus, a concept proposed seventy-five years ago by ROY and SHERRINGTON [20]
appears to have been substantiated.

The physiological competence of the circle of Willis and other anastomoses has
been demonstrated in one small series of measurements of cerebral blood flow before
and after acute occlusion of one carotid [21]. In relatively young individuals there
was no diminution in over-all cerebral perfusion, whereas in the presence of arteriosclerosis the adjustment was inadequate. The effects of gravitational stress [22] and
of variations in blood viscosity [23] and intracranial pressure [24] on cerebral blood
flow and vascular resistance have also been clarified.

The pharmacology of the cerebral circulation has been largely rewritten in terms
of the studies in man [25]. This is understandable since only in human subjects or
clinical disorders can relevant dosages of drugs be used, therapeutic effects studied
and possible species differences circumvented. These studies have served to reinforce
the characterization of a few of the cerebral dilator drugs but to question the efficacy
of many more. They may be expected to play a significant role in the evaluation of
new drugs and therapeutic procedures. The marked increases in cerebral blood flow
produced by the inhalation of carbon dioxide in 5 or 10% concentration has been
repeatedly demonstrated not only in normal man but in patients suffering from hypertension, cerebrovascular disease and a variety of disorders. The inhalation of high
concentrations of oxygen is accompanied by a slight cerebral vasoconstriction [19],
which may have resulted from the associated slight hyperventilation and hypocapnia.
Papaverine [26] and acetazolamide [27] are two of the very few drugs which have
been shown to be effective in increasing cerebral blood flow, while other drugs
formerly presumed to be cerebral vasodilators have been found to be ineffective
(e.g. nicotinic acid [28], alcohol [29]) or to produce significant vasoconstriction
(e.g. aminophylline [30] and other xanthines). Noradrenaline has been found to
constrict cerebral vessels as it does other vascular beds while adrenaline, with no
constrictor effect on cerebral vessels, permits an increase in cerebral blood flow *pari
passu* with a rise in blood pressure [31].

One of the unique values of methods for the study of the human cerebral circulation is the opportunity thus afforded for examining this dynamic function in the
pathogenesis of clinical diseases for which there are no entirely satisfactory animal
models. In essential hypertension [32] there is an increase in cerebrovascular resistance
which is exactly comparable to the increased mean arterial blood pressure so that
cerebral blood flow is not affected. This cerebral vasoconstriction does not appear
to be mediated by the sympathetic supply to the head, at least it is unaffected by the
blockade of both stellate ganglia in hypertensive patients [17]. These findings would
be compatible equally with the presence of a uniformly vasoconstricting substance
circulating in essential hypertension, or with an effective autoregulation in the cerebral
circulation. In selected groups of elderly individuals one finds an increasing incidence
of hypertension, cerebral atherosclerosis and mental deficit correlated with an increased cerebrovascular resistance, decreased cerebral blood flow and decreased

cerebral oxygen consumption, suggesting the operation of these factors in sequence to produce the mental changes [33].

A question of some importance in cerebrovascular disorders is the relationship of arterial blood pressure to the adequacy of cerebral blood flow in the presence of vascular disease which may confound the normal physiology. In uncomplicated essential hypertension an acute reduction of the arterial blood pressure brought about by sympathetic blockade of the trunk and lower extremities is accompanied by an almost comparable relaxation of cerebrovascular resistance, tending to maintain cerebral blood flow at its previous value [34] and suggesting that it is autoregulation which is responsible for the increased cerebrovascular resistance in essential hypertension. Such observations speak for the safety as well as the desirability of using drugs or procedures to reduce the blood pressure toward normal levels in uncomplicated essential hypertension. In the presence of cerebral arteriosclerosis, on the other hand, the situation is more complex. Where the resistance in one or more cerebral arteries is severe enough, it may become the limiting factor rather than arteriolar resistance in the perfusion of its vascular bed. Under those conditions the normal autoregulation is lost and blood flow to the affected region may become highly dependent on arterial blood pressure. In such individuals reduction in blood pressure by sympathetic blockade is followed by a marked fall in cerebral blood flow, a decrease in cerebral venous pO_2 and symptoms of cerebral ischemia [34]. It is reasonable, on the basis of such evidence, to postulate the existence of a similar mechanism rather than cerebrovascular spasm in many types of transient cerebral ischemic episodes associated with intermittent falls in arterial blood pressure. A question can be raised in such patients regarding the appropriateness of reducing a hypertension which may be necessary to maintain a minimal perfusion to some significant area. It has been suggested [35] that the increase in cerebral infarction and the decrease in hemorrhage reported from several countries over the past several years may have resulted from the introduction of hypotensive drugs and their use in individuals whose hypertension is complicated by cerebral arteriosclerosis.

Where stenosis of a major artery is the limiting factor in cerebral ischemia, its surgical correction offers a rational treatment. It is surprising that a definitive study has not yet appeared on the effects of such procedures on over-all cerebral blood flow. A cruder but perhaps useful pre-and post-operative diagnostic test would be the determination of the effects of temporary occlusion of each carotid separately on arterial-internal jugular oxygen differences.

In spite of whatever contributions they may have made to normal and abnormal human physiology, measurements of the total cerebral circulation have been of only limited value in the study of the individual patient with cerebrovascular disease. For diagnosis, localization and evaluation of progress of the disorder in the individual patient, we are still in need of suitable methods.

A simple, nontraumatic, yet reliable way of measuring cerebral blood flow in large numbers of patients and in the same patient many times would be of great clinical value. The approximation possible from head counts after the inhalation of a radioactive gas [36] may become sufficiently standardized and evaluated to permit it to fulfill all of these requirements. Continuous recording of changes in cranial blood volume by impedance measurement [37] may be shown to be a useful empirical procedure, and in the circumstances where it can be shown to be correlated with cerebral blood flow, perhaps a rough measure of it.

Methods for measuring regional circulation in the brain are necessary for diagnostic localization and for following the course of a small but clinically significant ischemic lesion where measurement of over-all cerebral blood flow is especially inappropriate since each region receives a weighting in the average which is determined by its blood flow. Here the regional clearances of radioactive gases, as developed by Ingvar and Lassen [38], offer great promise especially as the procedures for collimation attain higher resolution or as the gamma ray scintillation camera becomes a practical possibility. Use of an isotope such as Kr^{74} with anihilation gamma rays at 180° in conjunction with time resolution coincindence counting is still in the future but should be capable of reliable localization. Arteriography, cine-arteriography and subtraction techniques are being applied to yield transit and washout times which hopefully will be related to blood flow [39] in specific vessels or areas.

Thus, clinically applicable measurement of cerebral blood flow has demonstrated during life many of the inferences of functional change derived from morphological studies, has made possible pharmacological studies of greater relevance to human disease, and offers promise, already partially fulfilled, of contributions to the diagnosis and longitudinal study of individual patients.

References

1. Kety, S. S., Schmidt, C. F.: The determination of cerebral blood flow in man by the use of nitrous oxide in low concentrations. Amer. J. Physiol. 143, 53—66 (1945).
2. — — Nitrous oxide method for the quantitative determination of cerebral blood flow in man: theory, procedure, and normal values. J. clin. Invest. 27, 476—483 (1948).
3. Scheinberg, P., Stead, E. A.: The cerebral blood flow in male subjects as measured by the nitrous oxide technique, normal values for blood flow, oxygen utilization, glucose utilization, and peripheral resistance, with observations on the effect of tilting and anxiety. J. clin. Invest. 28, 1163—1171 (1949).
4. Bernsmeier, A., Siemons, K.: Die Messung der Hirndurchblutung mit der Stickoxydul-methode. Pflügers Arch. ges. Physiol. 258, 149—162 (1953).
5. Aizawa, T.: Problems of cerebral circulation. Tokyo: Kanehara Co. 1956.
6. Lassen, N. A., Munck, O.: The cerebral blood flow in man determined by the use of radioactive krypton. Acta physiol. scand. 33, 30—49 (1955).
7. Gottstein, U.: Der Hirnkreislauf unter dem Einfluß vasoaktiver Substanzen. Einzel-darst. theoret. klin. Med. 15, 11—186 (1962).
8. McHenry, L. C., Jr.: Quantitative cerebral blood flow determination, application of krypton-85 desaturation technique in man. Neurology (Minneap.) 14, 785—793 (1964).
9. Meyer, J. S., Gotch, F., Tomito, U., Akiyama, M.: New techniques for recording cerebral blood flow and metabolism in subjects with cerebrovascular disease. In: Cerebral vascular diseases. Transactions of the 5th Conference, 1966. (Siekert, R. G., Whisnant, J. P., Eds.) New York: Grune and Stratton (in press).
10. Lewis, B. M., Sokoloff, L., Wechsler, R. L., Wentz, W. B., Kety, S. S.: A method for continuous measurement of cerebral blood flow in man by means of radioactive krypton (Kr-79). J. clin. Invest. 39, 707—716 (1960).
11. Hill, L.: The physiology and pathology of the cerebral circulation: An experimental research. London: J. & A. Churchill 1896.
12. Kety, S. S.: Circulation and metabolism of the human brain in health and disease. Amer. J. Med. 8, 205—217 (1950).
13. Lassen, N. A.: Cerebral blood flow and oxygen consumption in man. Physiol. Rev. 39, 183—238 (1959).

14. CHOROBSKI, J., PENFIELD, W.: Cerebral vasodilator nerves and their pathway from the medulla oblongata. Arch. Neurol. Psychiat. (Chic.) **28**, 1257 (1932).
15. KETY, S. S.: Stellate ganglion blockade and the cerebral circulation (an editorial). Anesthesiology **20**, 697 (1959).
16. FORBES, H. S., COBB, S.: Vasomotor control of cerebral vessels. Res. Publ. Ass. nerv. ment. Dis. **18**, 201—217 (1938).
17. HARMEL, M. H., HAFKENSCHIEL, J. H., AUSTIN, G. M., CRUMPTON, C. W., KETY, S. S.: The effect of bilateral stellate ganglion block on the cerebral circulation in normotensive and hypertensive patients. J. clin. Invest. **28**, 415—418 (1949).
18. LINDEN, L.: The effect of stellate ganglion block on cerebral circulation in cerebrovascular accidents. Acta med. scand., Suppl. 301, **151**, 1 (1955).
19. KETY, S. S., SCHMIDT, C. F.: Effects of altered arterial tensions of carbon dioxide and oxygen on cerebral blood flow and cerebral oxygen consumption of normal young men. J. clin. Invest. **27**, 484—492 (1948).
20. ROY, C. S., SHERRINGTON, C. S.: On the regulation of the blood supply of the brain. J. Physiol. (Lond.) **11**, 85 (1890).
21. SHENKIN, H. A., CABIESES, F., VAN DEN NOORDT, G., SAYERS, P., COPPERMAN, R.: The hemodynamic effect of unilateral carotid ligation on the cerebral circulation of man. J. Neurosurg. **8**, 38 (1951).
22. HENRY, J. P., GAUER, O. H., KETY, S. S., KRAMER, K.: Factors maintaining cerebral circulation during gravitational stress. J. clin. Invest. **30**, 292—300 (1951).
23. NELSON, D., FAZEKAS, J. F.: Cerebral blood flow in polycythemia vera. Arch. intern. Med. **98**, 328—331 (1956).
24. KETY, S. S., SHENKIN, H. A., SCHMIDT, C. F.: Effects of increased intracranial pressure on cerebral circulatory functions in man. J. clin. Invest. **27**, 493—499 (1948).
25. SOKOLOFF, L.: The action of drugs on the cerebral circulation. Pharmacol. Rev. **11**, 1—85 (1959).
26. JAYNE, H. W., SCHEINBERG, P., RICH, M., BELLE, M. S.: The effect of intravenous papaverine hydrochloride on the cerebral circulation. J. clin. Invest. **31**, 111—114 (1952).
27. MITHOEFER, J. C., MAYER, P. W., STOCKS, J. F.: Effect of carbonic anhydrase inhibition on the cerebral circulation of the anesthetized dog. Fed. Proc. **16**, 88—89 (1957).
28. SCHEINBERG, P.: The effect of nicotinic acid on the cerebral circulation with observations on extracerebral contamination of cerebral venous blood in the nitrous oxide procedure for cerebral blood flow. Circulation **1**, 1148—1154 (1950).
29. BATTEY, L. L., HEYMAN, A., PATTERSON, J. L., JR.: Effects of ethyl alcohol on cerebral blood flow and metabolism. J. Amer. med. Ass. **152**, 6—10 (1953).
30. WECHSLER, R. L., KLEISS, L. M., KETY, S. S.: The effects of intravenously administered aminophylline on cerebral circulation and metabolism in man. J. clin. Invest. **29**, 28—30 (1950).
31. KING, B. D., SOKOLOFF, L., WECHSLER, R. L.: The effects of 1-epinephrine and 1-norepinephrine upon cerebral circulation and metabolism in man. J. clin. Invest. **31**, 273 to 279 (1952).
32. KETY, S. S., HAFKENSCHIEL, J. H., JEFFERS, W. A., LEOPOLD, I. H., SHENKIN, H. A.: Blood flow, vascular resistance and oxygen consumption of the brain in essential hypertension. J. clin. Invest. **27**, 511—514 (1948).
33. DASTUR, D. K., LANE, M. H., HANSEN, D. B., KETY, S. S., BUTLER, R. N., PERLIN, S., SOKOLOFF, L.: Effects of aging on cerebral circulation and metabolism in man. In: Human aging: A biological and behavioral study, pp. 59—76. Washington, D. C.: Public Health Service Publications No. 986, U. S. Govt. Printing Office 1963.
34. KETY, S. S., KING, B. D., HORVATH, S. M., JEFFERS, W. A., HAFKENSCHIEL, J. H.: The effects of an acute reduction in blood pressure by means of differential spinal sympathetic block on the cerebral circulation of hypertensive patients. J. clin. Invest. **29**, 402—407 (1950).
35. HOOBLER, S.: Discussion. In: Cerebral vascular diseases, Transactions of the 5th Conference, 1966. (SIEKERT, R. G., WHISNANT, J. P., Eds.) New York: Grune and Stratton (in press).

36. Veall, N., Mallett, B. L.: The two-compartment model using Xe^{133} inhalation and external counting. Acta neurol. scand. 41, Suppl. 14, 83—84 (1965).
37. Ziemnowicz-Radvan, S. A.: Rheographic regional method for evaluation of cerebral and ocular circulation in cardiac and cerebrovascular disease. J. Amer. Geriat. Soc. 13, 35—43 (1965).
38. Lassen, N. A., Hoedt-Rasmussen, K., Sorgense, S. C., Skinhoj, E., Cronquist, S., Bodforss, B., Ingvar, D. H.: Regional cerebral blood flow in man, determined by krypton-85. Neurology (Minneap.) 13, 719—727 (1963).
39. Hilal, S. K.: Densitometric evaluation of cerebral angiograms. In: Cerebral vascular diseases, Transactions of the 5th Conference, 1966. (Siekert, R. G., Whisnant, J. P., Eds.) New York: Grune and Stratton (in press).

The Focal Cerebrovascular Lesion.
A Survey of Recent Studies with Isotope Techniques
and their Clinical Implications*

D. H. INGVAR

I. Introduction

Cerebrovascular disorders constitute about two thirds of the clinical neurological clientele. This figure may increase even more with increasing longevity.

Clinical studies on cerebrovascular disorders have long since been dominated by patho-anatomical investigations, and, in addition, for the last decades, neuroradiological studies. Theories concerning the pathophysiology of apoplexy in its various forms have, therefore, by and large had a morphological basis, a fact pervading the vast literature in the field. However, several features of stroke syndromes—especially their non-fatal forms—, such as their often transient nature, the frequent lack of x-ray findings etc., cannot adequately be explained by present patho-anatomical knowledge. In addition, the morphological viewpoint has not proved very fruitful for the development of new forms of therapy for apoplexy.

Theoretical and technical advances in the last decade, which have brought quantitative techniques into clinical research, have also opened new roads for the study of cerebrovascular disorders. In the present survey, an account will be given of new clinical methods for measurements of the circulation in circumscribed parts of the brain. These methods have revealed several new aspects of the pathophysiology of occlusive cerebrovascular lesions, and this appears to have certain direct implications for the diagnosis and therapy of strokes.

II. Present Clinical Methods to Measure the Cerebral Blood Flow

The Kety-Schmidt method [1] measures the *total average* cerebral blood flow (and oxygen uptake). For cerebrovascular disorders, which mainly affect a circumscribed part of the brain, this method can only give limited information. It has proved of value for measurements of the global reduction of the cerebral blood flow and metabolism which follows a stroke and which usually is accompanied by a proportional intellectual impairment (dementia; [2]).

Further development of the Kety principles [3] has yielded methods for measurement of *regional* blood flow in the brain by external recording of the clearance of radioactive gamma emitting tracers (usually [133]Xenon), injected into the internal carotid artery [4, 5, 6, 7, 8, 9].

* Department of Clinical Neurophysiology, University Hospital, Lund, Sweden.

For the understanding of isotope clearance methods, the concept of the *mean transit time* (*t*) of the indicator particles plays a fundamental role [6, 7, 10]. When a freely diffusible indicator bolus is injected into the arterial inflow of an organ

$$\bar{t} = \frac{V}{F} \, , \tag{1}$$

where V is the volume of the organ, and F the total flow. However, if V and F are divided by W, the tissue weight,

$$\bar{t} = \frac{V/W}{F/W} \, , \tag{2}$$

which means that t becomes equal to the "distribution volume" (V/W) of the indicator, which is λ (the blood/tissue partition coefficient), divided by F/W, i.e. flow per unit weight, or f, the parameter wanted. Thus

$$\bar{t} = \frac{\lambda}{f} \, . \tag{3}$$

However, when the indicator transit is recorded over an organ by external counting following a sudden intra-arterial injection of an indicator bolus, a classical clearance curve is obtained, for which it holds that

$$\bar{t} = \frac{A}{H} \, , \tag{4}$$

where A is the area under the clearance curve (i.e. the integral of all different transit times), and H the initial height of the clearance curve, i.e. the total number of indicator particles to be cleared by the tissue under the detector. Both H and A are easily determined in counts per minute, respectively total counts, from the clearance curve. From equations (3) and (4) it follows that

$$f = \frac{\lambda \cdot H}{A} \, \text{ml/g/min} \, . \tag{5}$$

This simple ("stochastic") "height-over-area" calculation of regional blood flow in an organ does not include any assumptions as to certain tissue compartments.

Another method to calculate regional blood flow from externally recorded clearance curves assumes the presence of mainly two, relatively homogenous compartments in the brain, the grey and the white matter [6, 7]. Each clearance curve is considered as a biexponential function on this basis, and it yields, apart from a mean flow, also the flow in the grey and the white matter, as well as the relative weights of these compartments. Such calculations have given several interesting results. Regional changes within the grey matter flow have, for example, been demonstrated during mental effort [11]. The limitations of the "compartmental" method should not, however, be overlooked in clinical measurements. Evidently, pathological brains do not contain only normal grey and white matter.

A third and very simple method to calculate regional cerebral blood flow (in relative terms) takes only the initial, almost rectilinear slope of the clearance curve during the first two minutes into account. The slope is used as an index of flow. For rapid and dynamic clinical studies this simplification has proved most useful [12].

The above described isotope methods following intra-arterial indicator injection, have recently been thoroughly discussed in relation to techniques, e.g. the gamma camera, which give a dynamic two-dimensional isotope distribution display in an organ [9]. It can at present be concluded that for quantitative blood flow measurements over and around focal brain lesions, as well as for dynamic flow studies during various physiological conditions, the isotope clearance methods used with multiple (eight or sixteen) detectors [13, 14], have so far proved superior to other systems.

Thermoelectric methods for *continuous* measurements of regional cerebral blood flow (in relative terms) have also been used in man [15, 16]. These methods require, however, surgical exposure of the brain surface or stereotaxic implantations, and they do not lend themselves to routine studies in stroke patients.

Here it might be added that other techniques for *total* brain blood flow measurements, such as those based upon the Stewart-Hamilton principle [17] are of limited value for the study of focal vascular lesions. Of greater interest are methods for measurement of indicator transit times through the brain [18, 19] since a correlation exists between this transit time and the cerebral blood flow [20]. Rheoencephalography, the measurement of transcranial impedance changes, has not, in the author's opinion, proved to be of any clinical value in this context.

To this technical section it must finally be added that cerebral angiography and brain scanning techniques have made important advances in recent years with respect to cerebrovascular disorders. This large and important field lies, however, outside the scope of this survey and the reader is referred to appropriate textbooks and reviews [21, 22].

III. The Normal Regulation of the Cerebral Blood Flow

It now appears possible to present a fairly detailed concept of the normal regulation of the cerebral blood flow, based upon quantitative experimental and clinical measurements of the cerebral blood flow under various physiological conditions. These studies have mainly been made by the isotope methods presented above.

The systemic *blood pressure* provides the driving force of the cerebral blood stream. However, as pointed out already in 1934 by FOG [23], the cerebral blood flow does not follow pressure changes passively, but remains constant due to the so called *autoregulation*, which implies an adaptation of the cerebro-vascular resistance by vasodilatation when the pressure falls, and vice versa. Thus, the cerebral blood flow is normally not influenced by variations in the systemic pressure, when e.g. the body position changes, during rest and physical work, during emotions etc.

The mechanism of autoregulation is not fully known. It appears that metabolic factors may play a primary role. A fall in pressure may thus lower the pH around the contractile elements of the vessels and make them dilate, and vice versa. "Myogenic" theories have also been advocated to explain the unique capacity of the cerebral vascular bed to avoid flow changes when the systemic pressure varies (cf. ref. 9 for a thorough discussion of this topic).

ROY and SHERRINGTON, in 1890 [24], were the first to present evidence that it is the metabolic activity of the nervous tissue proper which sets its rate of blood flow. Ever since, the "intrinsic" mechanisms behind the "*metabolic control*" of the brain blood flow have been the object of many studies. It appears now that the oxygen consumption of the nervous tissue and its carbon dioxide output, plays a primary role in this regulation. Both in man and in animals, there is a very close coupling

9*

between the oxygen uptake and the functional activity (EEG) of the nervous tissue, and its blood flow [25, 26, 27, 28, 29]. However, the mode of action of carbon dioxide, possibly via the extracellular pH at different sites of the cerebral vascular bed, is still obscure. At steady state, in chronic disturbances of the acid-base balance, the extracellular (CSF) pH correlates better to the flow than does the arterial carbon dioxide tension [9].

Older and recent histological studies have proved that the cerebral vessels are equipped with abundant *vasomotor nerves*, including both adrenergic and cholinergic fibres [30, 31]. Apparently, there are important differences in vasomotor nerve supply between the "extracerebral" cerebral vessels and the "intracerebral" ones [32]. Thus, the large arteries at the base of the brain and around the circle of Willis are very richly supplied, while intraparenchymatous vessels of arteriolar size, or smaller, may only be equipped with single fibres, if any at all [33].

Clinical and vast amounts of experimental studies have so far failed to demonstrate that the cerebral vasomotor nerves—or for that matter the vagi and sinus nerves—serve any important function for the normal cerebral circulation. Stimulation and elimination of the sympathetic and parasympathetic components appears to change the normal flow only marginally (10 to 15%; [9]). Recently, however, evidence has been presented that vagal and sinus nerve impulses may alter the carbon dioxide sensitivity of the cerebral vessels [53]. It also remains an open question whether vasomotor impulses under pathological conditions (e.g. during "vasospasm"), or during abnormalities of the humoral transmission, may influence the cerebral blood flow critically.

Concerning *humoral factors* influencing the cerebral blood flow, none, apart from carbon dioxide, seem of importance under normal conditions. As was shown a long time ago [34], the carbon dioxide tension of the arterial blood exerts a dilator action on the cerebral vessels. The carbon dioxide of the blood apparently interacts with the "intrinsically" generated carbon dioxide (see above) to set the blood flow of the nervous tissue at an adequate level. This interaction is, however, not fully understood. It seems possible that the carbon dioxide of the arterial blood, and that generated by tissue metabolism, act upon different segments of the cerebral vascular bed (cf. [9]).

IV. The Occlusive Cerebrovascular Lesion

The etiology of various forms of occlusive cerebrovascular lesions—this much debated and still obscure topic—will not be discussed here. From the pathophysiological point of view, all such lesions, whether due to an embolus, to thrombosis, or haemorrhage, imply a more or less widespread, more or less long-lasting, focal cerebral ischemia (tissue hypoxia/anoxia). Although the full sequence of events following occlusive lesions are not known in detail yet, it appears justified, on the basis of neuroradiological data [35] to assume that a great number of such lesions are transient. This would mean that the microemboli, which are commonly held responsible for the majority of cerebral ischemic lesions, would sooner or later disappear, possibly due to thrombolysis. Thus, the flow (pressure-head) may be more or less completely re-established within the previously ischemic region, but often leaving a more or less permanent functional defect behind. Evidence for this view has been obtained from stroke cases in which serial angiograms have been made [36].

Focal cerebral ischemia (tissue hypoxia/anoxia), even of short duration for some minutes only, completely upsets the normal regulatory factors reviewed above. The high metabolism of nervous tissue requires a constant and high oxygen and glucose supply. If this is suddenly diminished, or ceases, a severe tissue lactacidosis develops rapidly, resulting in a lowering of the tissue pH, a partial or complete depression of the oxidative metabolism, and a number of changes in the intermediary metabolism[9]. In permanent ischemia, tissue necrosis follows. Provided the lesion is large enough, or superficial, several of the biochemical changes mentioned, such as the lactacidosis, can be followed also in the cerebrospinal fluid [9].

However, if, as apparently happens quite often, a certain pressure-head is re-established within the once ischemic region, the flow in this region shows several abnormal features which, in part, can be traced by angiography and by flow measurements for days and even weeks.

Firstly, *the "metabolic control" disappears* [29]. There is no longer a coupling between the functional activity (EEG) of the tissue with its depressed metabolism, and its blood flow. The vascular bed appears "vasoparalytic" due to the acidosis, and—at normal systemic pressure—the flow may by far exceed the metabolic demands ("luxury perfusion"; LASSEN [37, 9]). The venous blood coming from such "re-perfused" previously ischemic regions may in fact have a bright red colour [9].

Secondly, the vasoparalysis also implies *a loss of autoregulation* and in the revascularized lesion area the flow passively follows changes of the systemic pressure [9, 29]. In this context, hypotension as a cause of strokes, mainly of the transient type, should be mentioned. According to this theory, cerebral arteriosclerosis might cause focal impairment of the autoregulation and, hence, a drop of blood pressure might lead to focal ischemia. However, dynamic flow studies (see below) do not reveal such impairments in patients with this form of repeated transient cerebral symptoms. These patients appear to have a normal autoregulation between the attacks [57]. This evidence would thus speak against hypotension—and indirectly favour the microembolic theory—as a cause of focal cerebral ischemia.

In this schematic description, it should be emphasized that several focal cerebrovascular lesions in all probability may contain a whole spectrum of the phenomena described. Thus, a large lesion may have a totally ischemic nucleus, which ultimately undergoes necrosis. In the surroundings, different forms of hypoxic/anoxic derangements of various severity can be encountered in which the metabolic and flow abnormalities described are more or less pronounced (Fig. 1) [38]. Furthermore, during the post-anoxic phase, with luxury perfusion, there are also permeability changes of various types with brain oedema and/or extravasation of blood, phenomena known since long in neuropathology [39]. It appears entirely likely that such changes in their turn for mechanical reasons, may jeopardize the flow even further. The relationship between brain oedema and the pathophysiological events described here will require further study.

V. Diagnostic Implications

The clinical diagnosis of a focal cerebrovascular lesion usually includes cerebral angiography by which in many cases an occlusion, or, in haemorrhages or focal edema, an expansive mass is demonstrated. However, cerebral angiography is in many stroke cases often completely negative, as emphasized above [35].

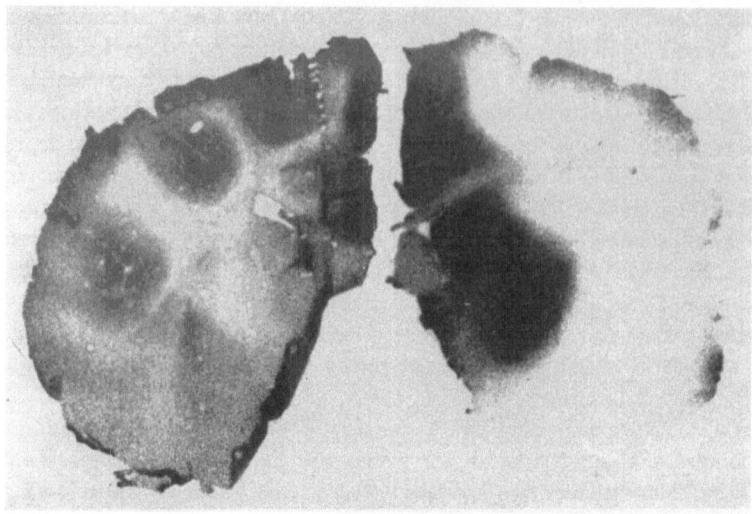

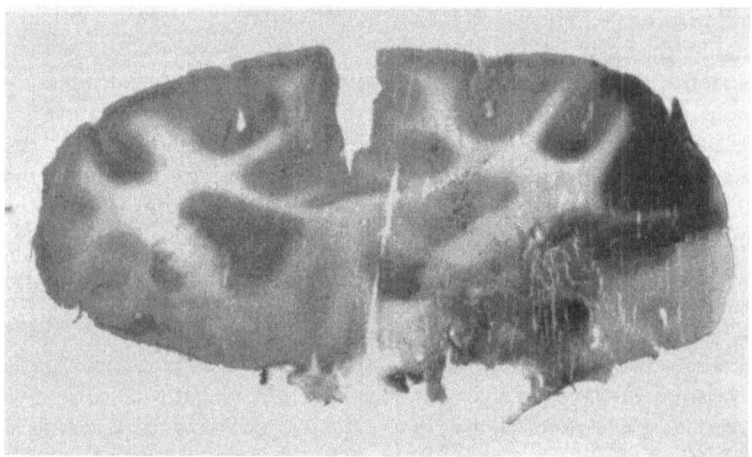

Fig. 1. *Cerebral blood flow in and around a focal ischemic lesion.*

Autoradiographs of frontal sections of cat brain, following infusion for 1 min of ^{14}C-antipyrine. The concentration of antipyrine is a function of the local blood flow rate. In the two experiments shown, the middle cerebral artery had been ligated under visual inspection.

Upper photograph: Here the middle cerebral artery was ligated and a period of hypotension followed for 20 min. Ninety min. later hypertension was produced for one hour by means of infusion of Norepinephrine (maximal blood pressure 200 mmHg). Two and a half hours following the ligation the animal was sacrificed. The autoradiograph shows a large ischemic area (white) on the side of the ligation which is surrounded by a hyperemic (black) zone.

Lower photograph: This animal was treated in the same way as that in the upper photo, but the ligature was released after 20 min. The flow in the middle cerebral artery territory then resumed. Fourty-eight hours later, hypertension was produced for 6 hours, the ^{14}V-antipyrine was infused, and the animal was sacrificed. The infarction (lower right corner) is smaller, and surrounding it a marked hyperemic zone is observed.

(Courtesy of Prof. C. FIESCHI, Dept. of Neurology, University Hospital, Genova).

Recently, CRONQVIST and LAROCHE [40] have pointed out that in the acute phase of (usually smaller) cerebrovascular lesions, subtle specific angiographic signs can be demonstrated which are sometimes the *only* indices of a local flow abnormality in the region corresponding to the focal symptoms of the patient.

In the region which has been exposed to transient ischemia, a circumscribed *contrast* "*blush*" may be seen and/or "*early*" *veins*, which signify a rapid contrast passage. These signs form the radiographic counter part to the "luxury perfusion" phenomenon, i.e. the postanoxic "vasoparalytic" focal flow derangement. If repeated angiographies are made, the signs mentioned are found to be transient, although in many cases the focal neurological deficits may persist undiminished.

Multiple detector measurements of cerebral blood flow in stroke cases [41, 42, 43] have also revealed patterns of focal flow abnormalities which coincide with the concepts presented above.

Thus, in the acute phase, prior to re-establishment of flow, or over large lesions, a region of ischemia, i. e. with subnormal flow can be found. However, in many cases, and often in the border-zone of an ischemic lesion, it has been possible to demonstrate focal cerebral *hyperemia*, i.e. "luxury perfusion". This may in certain cases exceed the normal range by twice or more. Gradually, the hyperemia subsides over the first week or two, following the acute onset of the lesion, as seen in repeated flow studies [36].

The focal or peri-focal hyperemia implies the presence of tissue compartments perfused at a rate much higher than normal. Isotope clearance curves from such regions show typical initial "peaks" indicating an abnormally rapid transit of certain portions of the indicator (Fig. 2) [43]. Here it should be added that the flow abnormalities described in this section are seen in their most clearcut form in superficial supratentorial lesions, usually within the middle cerebral artery territory. Many small and deep vascular lesions, e.g. of the brain stem, cannot be measured by the isotope techniques described. Only if such lesions involve critical structure of the brain stem may they give rise to a reduction of the total function and the blood flow of the brain [44].

Finally, in "chronic" cerebrovascular cases, e.g. in hemiplegic patients studied several months after the acute stroke, the cerebral blood flow shows no marked focal variations, but instead, a general over-all reduction. In studies with the Kety technique [2], as well as in regional flow studies [42], it has been found that this flow reduction correlates with the intellectual impairment (dementia) of these patients. Here, it should be mentioned that the relationship between senile dementia and chronic cerebrovascular disorder is still unclear. Many severely demented senile patients without history of strokes also show marked cerebral flow reductions [45].

VI. Dynamic Studies of the Blood Flow within the Cerebrovascular Lesion

A number of procedures are presently being developed in order to bring out latent focal abnormalities of the cerebral blood flow. The aim is to measure the marginal capacity of the cerebral circulation to endure functional loads of different types. Stroke patients with symptoms of intermittent cerebral ischemia are of greatest interest in this respect.

Case 6: middle cerebral artery occlusion

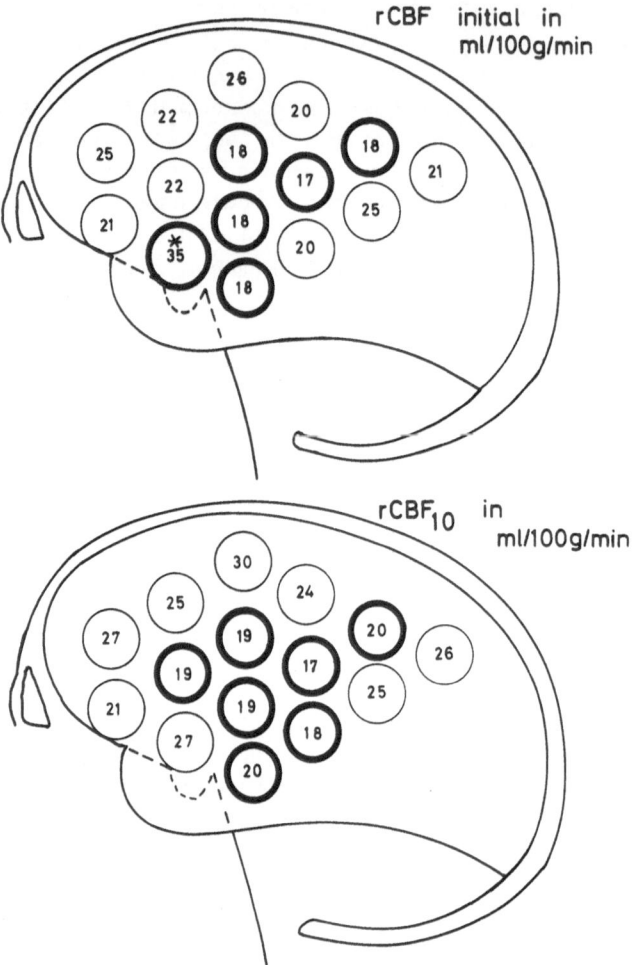

Fig. 2. *Measurement of regional cerebral blood flow in a 78 year old woman who 1 day earlier had suffered a complete right-sided hemiparesis with aphasia.* The patient had no history of previous epilectic attacks, and was not hypertensive. Angiography showed an occlusion of the left middle cerebral artery. Brain scanning was normal. A spinal tap showed normal CSF. The patient died one week following the study.

The upper diagram shows the flow values, calculated from the initial slope of the clearance curves (p. 137) in 16 regions, distributed over the left hemisphere as shown. Note low values in the middle of the lesion, and a high value (35 ml/100 g/min) in the border zone in the Sylvian region. The flow values outside the lesion showed subnormal values also.

The lower diagram shows examples of flow curves from the focus, the hyperemic perifocal area, and from "normal" parts of the hemisphere studied. Note the "peaked" clearance curve in the peri-focal area, indicating the presence of rapidly perfused tissue compartments ("luxury perfusion"; see text).

(Courtesy of Dr. OLAF POULSEN, Dept. of Clinical Physiology, Bispebjerg Hospital, Copenhagen).

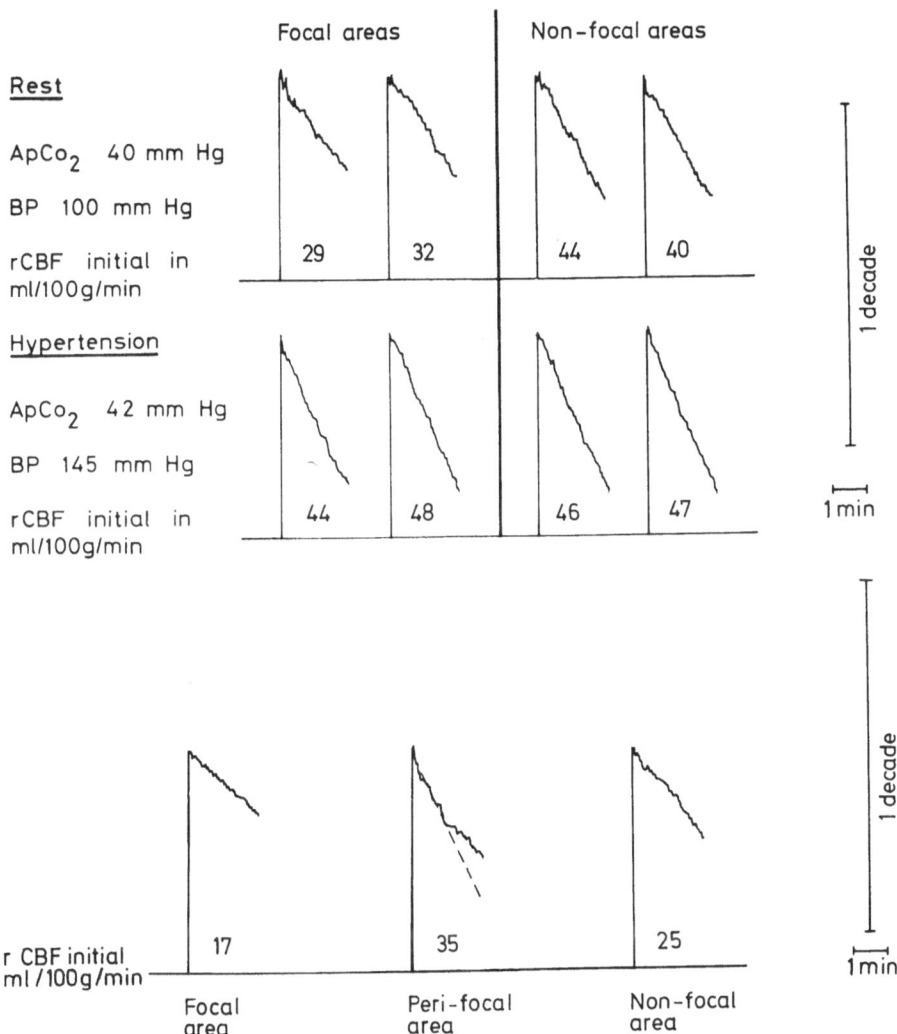

Fig. 3. *Demonstration of loss of autoregulation in an ischemic cerebral lesion.*

The patient was a 68 year old man without history of hypertension or cardiac disease. In October 1967 he woke up one morning with an almost complete right-sided hemiparesis and aphasia. During the following month the symptoms disappeared to some extent. Angiography showed an occlusion of the left middle cerebral artery, and a brain scan showed an increased uptake in the left temporal region. Lumbar puncture 11 days following the onset of symptoms showed hemorrhagic CSF. EEG showed a slow wave focus in the left temporal region.

Measurement of regional cerebral blood flow 9 days after the onset of symptoms showed low values in the center of the lesion, and, in one channel, a "peaked" clearance curve (cf. Fig. 2). During a period of hypertension (from 100 mmHg to 145 mmHg), induced by Aramine, it was found that the autoregulation was preserved in non-focal areas (right four curves), while within the focus the flow passively followed the increase in pressure (left four clearance curves). Also in this case the flow was calculated by means of the initial slope method (see Fig. 3).

(Courtesy of Dr. OLAF POULSEN, Dept. of Clinical Physiology, Bispebjerg Hospital, Copenhagen).

As pointed out above, transient regional cerebral hypoxia/anoxia gives rise to a long-lasting diminution, or complete loss of autoregulation. This abnormality has recently been studied experimentally, and in acute apoplectics, by several groups [29, 38, 43]. Thus, transient hypertension, induced by *r.g.* Aramine, does not change the flow in normal parts of the brain, but augments the flow in regions where autoregulation is lost (Fig. 3).

A similar differentiation between normal and latently abnormal parts of the cerebral vascular bed, may also be traced during repeated angiographies, when the dilatation due to the contrast material might be more pronounced in the focal area during the second procedure [9].

Carbon dioxide inhalation has also been used in order to enhance focal flow abnormalities. There is, as mentioned, some evidence which, however, cannot be considered unequivocal, that carbon dioxide and "pressure" act upon different "receptors" in the cerebral vessels. The abnormalities of autoregulation and of carbon dioxide sensitivity may thus differ in various parts of a vascular lesion, an interesting finding which, however, merits further study [38].

A "neuronal" type of activation consists of letting the patient carry out some form of increased mental activity. In normals, mental effort augments (or diminishes) the grey matter blood flow in various regions of the hemisphere [11] and may give rise to specific patterns of changes in regional cerebral blood volume [46]. A limited experience in patients with brain lesions, including cerebrovascular ones, suggests that a combination of mental tests and regional brain blood flow (or blood volume) measurements may be used clinically to measure the intellectual and cerebral circulatory sequelae of strokes [11].

VII. EEG and Apoplexy

As mentioned, there is normally a correlation between the frequency content of the EEG—in adult man, awake—and the cerebral blood flow [25, 26, 27]. Also in cerebral lesions which have passed the first month following the acute onset of symptoms, this correlation is present. The average frequency of the EEG (in awake adults) can thus be used to some extent to predict the cerebral blood flow. A limited experience from chronic focal lesions shows that the prediction of flow from the EEG also seems to hold on the regional plane [47]. For deeper lesions with projected EEG abnormalities, the situation is less clear, however.

In the acute phase of an ischemic cerebral lesion which has given rise to a metabolic depression and "luxury perfusion", the normal EEG/flow correlation is lost [29]. In this state, the slowing of the EEG over the lesion probably still retains its relation to the regional oxygen uptake, *i.e.* the EEG slowing mirrors the focal reduction of the cerebral oxidative metabolism.

Finally, it is well-known that certain occlusive cerebrovascular lesions, especially the embolic ones, may, apart from a focal EEG slowing, also give rise to focal epileptic discharges. Although the nature of epileptic processes still on the whole remains obscure, there is reason to believe that acute electrolyte disturbances caused by the tissue anoxia, which is known to lead to massive neuronal depolarisation and potassium release, may underly such paroxysmal electrical phenomena which—in accordance with the concepts presented above—might be termed "luxury EEG" [48, 49].

VIII. Therapeutic Considerations

Lack of oxygen within the nervous tissue may even after 3 to 5 minutes lead to permanent damage and focal symptoms. The rapidity with which this damage sets in, makes it unlikely that an effective and practicable causal therapy for occlusive lesions can be developed.

The only form of causal therapy in strokes used with some success is hyperbaric oxygen, administered in a pressure tank. The few cases so far treated, indicate that this therapy, if started early enough, may keep a limited ischemic cerebral region sufficiently oxygenated until revascularisation takes place [50, 51].

The demonstration of focal hyperemia, or "luxury perfusion", in the post-ischemic phase of an occlusive lesion, has radically changed the classical concepts which so far have dominated the therapeutic endeavours in acute apoplexy.

In strokes, numerous vasodilator agents have been tried in the belief that the blood flow in the focus was insufficient. As shown above, this concept is probably by and large erroneous, since apparently a great number of the occlusive processes disappear, and are followed by a massive vasodilatation and hyperemia in the period of revascularisation.

It appears in fact contraindicated to administer cerebral vasodilatants to stroke cases in the acute phase, due to what is now called the "*intracerebral-steal*" phenomenon [9, 52, 53]. This implies that a massive dilatation of the normal cerebral vascular bed by *e.g.* carbon dioxide, may divert the flow *from* the lesion area, by stealing pressure-head from the vasoparalytic focus. This sequence of events has in fact been shown to take place in patients, *i.e.* that carbon dioxide inhalation may diminish the flow in an ischemic focus [43].

On the other hand, the vasoparalytic focus may be benefited by the opposite, a *vasoconstriction* of the normal brain vessels. (This has been termed the "Robin Hood"-phenomenon, since it implies taking flow from the rich tissue and giving it to the poor [9]). There is then a diversion of flow from the constricted normal areas to the vasoparalytic focal region. The dramatic beneficial effects of administration of theophylline—a vasoconstrictor for the normal brain—in certain acute stroke cases, may have this background [54]. A systematic use of hyperventilation—the strongest known cerebral vasoconstricting measure—has also been tried successfully to diminish the effects of focal or general cerebral hypoxia [55, 56].

IX. Summary

1. Clinical studies in recent years with isotope techniques for measurements of *regional* cerebral blood flow have revealed new aspects of the pathophysiology of occlusive cerebrovascular lesions.

2. In focal hypoxic/anoxic lesions of the brain, the normal "metabolic control" of the cerebral blood flow is destroyed and the autoregulation of the cerebrovascular resistance is impaired or disappears. The flow in the focus then becomes pressure passive, and—at normal systemic blood pressure—supernormal ("luxury perfusion").

3. The "luxury perfusion" can actually in some cases be seen in the angiogram in the acute phase of a stroke as a focal contrast "blush" or as "early" veins. Flow measurements over such a focus discloses typical "peaked" clearance curves which yield supernormal flow values.

4. Over superficial occlusive lesions a slow EEG pattern develops which reflects the metabolic depression in the region. Once the acute phase is over (about 2 weeks), the normal correlation between EEG frequency content and cerebral blood flow returns. Projected EEG abnormalities following deep lesions cannot be interpreted according to such simple rules.

5. At present, "dynamic" procedures are being developed in order to bring out latent regional abnormalities of the cerebral circulation, especially in patients with symptoms of intermittent cerebral ischemia. Thus, induced hypertension may reveal an area with loss of autoregulation. A combination of psychological tests and measurements of regional cerebral blood flow (and blood volume) may possibly also be used to measure the intellectual and cerebral circulatory deficit in stroke patients.

6. The "luxury perfusion" in and around an acute cerebrovascular lesion provides a contraindication for the use of drugs with a cerebral vasodilatant action (*e.g.* carbon dioxide). They may lead to "intracerebral steal", *i.e.* diversion of flow *from* the vasoparalytic lesion area. On the other hand, vasoconstriction of normal parts of the cerebral vascular bed, e.g. by hyperventilation, may divert the flow from the normal brain *to* the vasoparalytic focus. Measures which cause cerebral vasoconstriction may at present therefore be considered to have a sounder rationale than vasodilatants in stroke therapy.

Acknowledgements: The author was supported by grants from the Swedish Medical Research Council, and by the Wallenberg Foundation, Stockholm. Erik Skinhöj, M. D., and N. A. Lassen, M. D., Bispebjerg Hospital, Copenhagen, kindly assisted the author with helpful criticism in preparing this review.

References

1. Kety, S. S., Schmidt, C. F.: Nitrous oxide method for quantitative determination of cerebral blood flow in man. J. clin. Invest. **27**, 476 (1948).
2. Heyman, A., Patterson, J. L., Duke, T. W., Battey, L. L.: The cerebral circulation and metabolism in arteriosclerotic hypertensive cerebrovascular disease. New Engl. J. Med. **249**, 223 (1953).
3. Kety, S. S.: In: McDowall, R. J. S. (Ed.). The control of the circulation of the blood. Suppl. vol., pp. 176—183. London: Dawson 1956.
4. Lassen, N. A., Høedt-Rasmussen, K., Sørensen, S. C., Skinhøj, E., Cronqvist, S., Bodforss, B., Ingvar, D. H.: Regional cerebral blood flow in man determined by krypton 85. Neurology (Minneap.) **13**, 719 (1963).
5. Ingvar, D. H., Lassen, N. A.: Methods for cerebral blood flow measurements in man. Brit. J. Anaesth. **37**, 216 (1965).
6. Høedt-Rasmussen, K., Sveinsdottir, E., Lassen, N. A.: Regional cerebral blood flow in man determined by intra-arterial injection of radioactive inert gas. Circulat. Res. **18**, 237 (1966).
7. — Regional cerebral blood flow. The intra-arterial injection method, p. 81. Copenhagen: Munksgaard 1967.
8. Ingvar, D. H., Lassen, N. A. (Eds.): Regional cerebral blood flow. An International Symposium, March 6, 1965. Copenhagen: Munksgaard 1965.
9. — —, Siesjö, B. K., Skinhøj, E.: Cerebral blood flow and cerebro-spinal fluid. III. International Symposium, Lund-Copenhagen, May 9—19, 1968. Scand. J. Lab. clin. Invest. Suppl. **102** (1968).
10. Zierler, K.: Two equations for measuring flow by external monitoring of radioisotopes. Circulat. Res. **16**, 309 (1965).
11. Ingvar, D. H., Risberg, J.: Increase of regional cerebral blood flow during mental effort in normals and in patients with focal brain disorders. Exp. Brain Res. **3**, 195 (1967).

12. PAULSEN, O. B., CRONQVIST, S., RISBERG, J., JEPPESEN, F.: Regional cerebral blood flow. A comparison of 8-detector and 16-detector equipments. J. nucl. Med. (1968) (in press).
13. INGVAR, D. H., LUNDMARK, T., RISBERG, J., VON SABSAY, E., BURKLINT, U., SUNDELIN, S.: Recording of multiple clearance curves by means of a magnetic core memory. Scand. J. Lab. clin. Invest. Suppl. 102 (1968), XI, H.
14. LASSEN, N. A.: Preliminary experience with oscilloscope and polaroid camera as recorder unit in a multichannel scintillation detector instrument. Scand. J. Lab. clin. Invest. Suppl. 102 (1968), XI, I.
15. COOPER, R., CROW, H. J., GREY WALTER, W., WINTER, A. L.: Regional control of cerebral vascular reactivity and oxygen supply in man. Brain Res. 3, 174 (1966).
16. BETZ, E., WÜLLENWEBER, R.: Fortlaufende Registrierung der lokalen Gehirndurchblutung mit Wärmeleitsonden am Menschen. Klin. Wschr. 40, 1056 (1962).
17. NYLIN, G., SILFVERSKIÖLD, B. P., LÖFSTEDT, S., BERGSTRÖM, O., KEDLUND, S.: Studies on cerebral blood flow in man, using radioactive labelled erythrocytes. Brain 83, 293 (1960).
18. OLDENDORFF, W. H., KITANO, M.: Isotope study of brain blood turnover in vascular disease. Arch. Neurol. Psychiat. (Chic.) 12, 30 (1965).
19. WILCKE, O.: Eine einfache Methode zur Bestimmung der Hirndurchblutung mit Radio-Isotopen. Acta neurochir. (Wien) 12, 31 (1964).
20. GREITZ, T., CRONQVIST, S.: Angiographic evaluation of cerebral circulation time and regional cerebral blood flow. A comparative study. Scand. J. Lab. clin. Invest. Suppl. 102 (1968), XI, A.
21. TAVERAS, J. M., WOON, E. H.: Diagnostic neuroradiology. Baltimore: William and Wilkins Company 1964.
22. AHLSTRÖM, S., BODFORSS, B., CRONQVIST, S.: Experiences with gamma-encephalography. Acta radiol. (Stockh.) 4, 417 (1966).
23. FOG, M.: Om piaarteriernas vasomotoriske reaktioner, p. 183. Copenhagen: Munksgaard 1934.
24. ROY, C. S., SHERRINGTON, C. S.: The regulation of the blood supply of the brain. J. Physiol. (Lond.) 11, 85 (1890).
25. INGVAR, D. H.: Cortical state of excitability and cortical circulation. In: Reticular formation of the brain, pp. 381—408. Henry Ford Hospital Symposium, Detroit, March 14—16, 1957. Boston: Little Brown 1958.
26. GLEICHMANN, U., INGVAR, D. H., LASSEN, N. A., LÜBBERS, D. W., SIESJO, B. K., THEWS, G.: Regional cerebral cortical metabolic rate of oxygen and carbon dioxide, related to the EEG in the anesthetized dog. Acta physiol. scand. 55, 82 (1962).
27. OBRIST, W. D., SOKOLOFF, L., LASSEN, N. A., LANCE, M. H., BULTER, R. N., FEINBERG, I.: Relation of EEG to cerebral blood flow and metabolism in old age. Electroenceph. clin. Neurophysiol. 15, 610 (1963).
28. BALDY-MOULINIER, M., INGVAR, D. H.: EEG frequency content related to regional blood flow of cerebral cortex in cat. Exp. Brain Res. 5, 55 (1968).
29. FREEMAN, J., INGVAR, D. H.: Elimination by hypoxia of cerebral blood flow autoregulation and EEG relationship. Exp. Brain Res. 5, 61 (1968).
30. PENFIELD, W.: Intracerebral vascular nerves. Arch. Neurol. Psychiat. (Chic.) 27, 30 (1932).
31. FALCK, B., MCHEDLISHVILI, G., OWMAN, C.: Histochemical demonstration of adrenergic nerves in cortex pia of rabbit. Acta pharmacol. (Kbh.) 23, 133 (1965).
32. MCHEDLISHVILI, G.: Vascular mechanisms pertaining to the intrinsic regulation of the cerebral circulation. Circulation 30, 597 (1964).
33. OWMAN, C., FALCK, B., NIELSEN, K.: Adrenergic innervation of the pial circulation. Scand. J. Lab. clin. Invest. Suppl. 102 (1968), VI, B.
34. GIBBS, F. A., GIBBS, E. L., LENNOX, W. G.: Changes in human cerebral blood flow consequent on alterations in blood gases. Amer. J. Physiol. 111, 557 (1936).
35. MARSHALL, J.: The management of cerebrovascular disease, 199 pp. London: Churchill 1965.
36. CRONQVIST, S., EKBERG, R., INGVAR, D. H., MÜLLER, R.: To be published.
37. LASSEN, N. A.: The luxury-perfusion syndrome and its possible relation to acute metabolic acidosis localised within the brain. Lancet 1966 II, 1113.

38. AGNOLI, A., FIESCHI, C., BOZZAO, L., BATTISTINI, N., PRENCIPE, M.: Autoregulation of cerebral blood flow: studies during drug induced hypertension in normal subjects and in patients with cerebrovascular diseases. Circulation **38**, 800 (1968).
39. LANGFITT, T. W., MARSHALL, W. J. S., KASSELL, N. F., SCHUTTA, H. S.: The pathophysiology of brain swelling produced by mechanical trauma and hypertension. Scand. J. Lab. clin. Invest. Suppl. **102** (1968), XIV, B.
40. CRONQVIST, S., LAROCHE, F.: Transitory "hyperaemia" in focal cerebral vascular lesions studied by angiography and regional cerebral blood flow measurement. Brit. J. Radiol. **40**, 270 (1967).
41. — EKBERG, R., INGVAR, D. H.: Regional cerebral blood flow related to neuroradiological findings. Acta neurol. scand. Suppl. **14**, 176 (1965).
42. INGVAR, D. H.: The pathophysiology of occlusive cerebrovascular disorders, related to neuroradiological findings, EEG, and measurements of regional cerebral blood flow. Acta neurol. scand. **43**, Suppl. 31, 93 (1967).
43. POULSEN, O.: Regional cerebral blood flow in apoplexy due to middle cerebral artery occlusion. To be published 1969.
44. INGVAR, D. H., HÄGGENDAL, E., NILSSON, N. J., SOURANDER, P., WICKBOM, I., LASSEN, N. A.: Cerebral circulation and metabolism in a comatose patient. Arch. Neurol. Psychiat. (Chic.) **11**, 13 (1964).
45. — OBRIST, W. D., CHIVIAN, E., CRONQVIST, S., RISBERG, J., GUSTAFSON, L., HÄGERDAL, M., WITTBOM-CIGÉN, G.: General and regional abnormalities of cerebral blood flow in senile and "presenile" dementia. Scand. J. Lab. clin. Invest. Suppl. **102** (1968), XII, B.
46. RISBERG, J., INGVAR, D. H.: Regional changes in cerebral blood volume during mental activity. Exp. Brain Res. **5**, 72 (1968).
47. SULG, I. A., INGVAR, D. H.: Regional cerebral blood flow and EEG in occlusions of the middle cerebral artery. Scand. J. Lab. clin. Invest. Suppl. **102** (1968), XVI, D.
48. HADJI-DIMO, A. A., CHRIST, R., HEIPERTZ, R., BROCK, M.: Acute electrocorticographic changes following localized cortical freezing in the cat. Preliminary report. Scand. J. Lab. clin. Invest. Suppl. **102** (1968), XIV, E.
49. GROSSMAN, R. G.: Intracellular potentials of motor cortex neurons in cerebral ischemia. Electroenceph. clin. Neurophysiol. **24**, 291 (1968).
50. INGVAR, D. H., LASSEN, N. A.: Treatment of focal cerebral ischemia with hyperbaric oxygen. Acta neurol. scand. **41**, 92 (1965).
51. SALTZMAN, H. A.: Rational normobaric and hyperbaric oxygen therapy. Ann. intern. Med. **67**, 843 (1967).
52. SYMON, L.: Experimental evidence for "intracerebral steal" following CO_2 inhalation. Scand. J. Lab. clin. Invest. Suppl. **102** (1968), XIII, A.
53. WÜLLENWEBER, R.: "Intracerebral steal" in man recorded by a heat clearance technique. Scand. J. Lab. clin. Invest. Suppl. **102** (1968), XIII, C.
54. OLIVARIUS, B. DE FINE: Apoplexia cerebri, p. 305. Copenhagen: Munksgaard, Thesis 1964.
55. GORDON, E., ROSSANDA, M.: The importance of the cerebrospinal fluid acid-base status in the treatment of unconscious patients with brain lesions. Scand. J. Lab. clin. Invest. Suppl. **102** (1968).
56. SOLOWAY, M., NADEL, W., ALBIN, M. S., WHITE, R. I.: The effect of hyperventilation on subsequent cerebral infarction. Anesthesiology **29** (1968) (in press).

Note added in proof: The present review was prepared in early 1968. Since then a number of papers have appeared concerning the use of isotope techniques in cerebrovascular disorders. The reader is referred to two larger publications for additional references:

BROCK, M., FIESCHI, E., INGVAR, D. H., LASSEN, N. A., SCHÜRMANN, K. (Eds): *Cerebral blood flow.* Clinical and experimental results. Papers presented at the International Symposium on Cerebral Blood Flow, 291 pp. Mainz, W. Germany, April 10—12, 1969. Springer Berlin-Heidelberg-New York, 1970.
LASSEN, N. A., INGVAR, D. H.: Radioisotopic assessment of regional cerebral blood flow. In: POTSCHEN, J. (Eds): *Progress in nuclear medicine.* New York: Karger, 1970. (In press).

The "Intracerebral Steal":
A Phenomenon in the Pathogenesis of Focal Brain Ischemia*

C. Fazio, C. Fieschi, and A. Agnoli

"Intracerebral steal" is a haemodynamic phenomenon widely reported in recent years by such authors as Fazio (1964), Feindell and Perot (1965), Loeb and Meyer (1965), Fazio et al. (1967), Shalit et al. (1967), Brawley et al. (1967), Lassen (1967), Symonds (1967), Fieschi et al. (1968), Paulson (1968).

A brief description of this phenomenon is as follows: when local vascular resistance decreases, a redistribution of the blood flow between various regions of the brain occurs; blood is "stolen" to the disadvantage of one area and to the advantage of others, thus producing an ischemia in the disadvantaged area. In most instances, however, one has to assume an impairment of autoregulation in the involved areas.

Intracerebral steal has been convincingly demonstrated only in pathological conditions. It was first surmised in patients with large arterio-venous malformations (Loeb and Favale, 1962); subsequently, the study of acute, focal brain lesions has provided a wealth of data. This report deals with the possible occurrence of intracerebral steal in previously asymptomatic patients, for whom, therefore, it may have constituted the primary pathology leading to focal brain ischemia.

This concept, then, suggests that in some cases of intracerebral steal, vascular supply to a particular area becomes deficient, not because of stenosis or occlusion of an afferent artery, but because of a primary shift of blood to other cerebral areas (Fazio, 1964).

The mechanisms responsible for such a shift require that the perfusion pressure in that area decreases below the lower limits compatible with an efficient cerebral autoregulation (Fieschi, 1967). This local drop in perfusion pressure—in order that the "steal" is properly called intracerebral—must be produced by hemodynamic phenomena taking place in other areas of the brain.

In cases of anatomic shunts, it is easily acceptable that the shunt itself or relatively small further reductions in pressure may diminish the perfusion pressure downstream, and induce ischemia of a brain area with otherwise normal autoregulation. Besides cases of arteriovenous malformations of fistulas, this may rarely happen in patients with presistent embryonic anastomoses between carotid and basilar arteries (Pastorino and Piola, 1966) and in other anomalies of the circle of Willis leading to the so-called "enlarged territory" (Fazio et al., 1965).

More commonly, however, the intracerebral steal becomes effective when the local autoregulation is already compromised by subtle changes in vascular reactivity (Mamo

* Istituto di Clinica delle Malattie Nervose e Mentali dell'Universita' di Roma (Direttore: Prof. C. Fazio).

et al., 1967). Indeed, it has been demonstrated that autoregulation of cerebral vessels is a delicate phenomenon, that in cases of overt or even latent brain pathology may be easily impaired uniformly throughout the brain, or more frequently, in some circumscribed areas more than in others (Agnoli e Bozzao, 1969).

Because of impaired autoregulation, adequate protection of cerebral tissues in the face of changes of perfusion pressures is not maintained. Thus, changes in cerebral

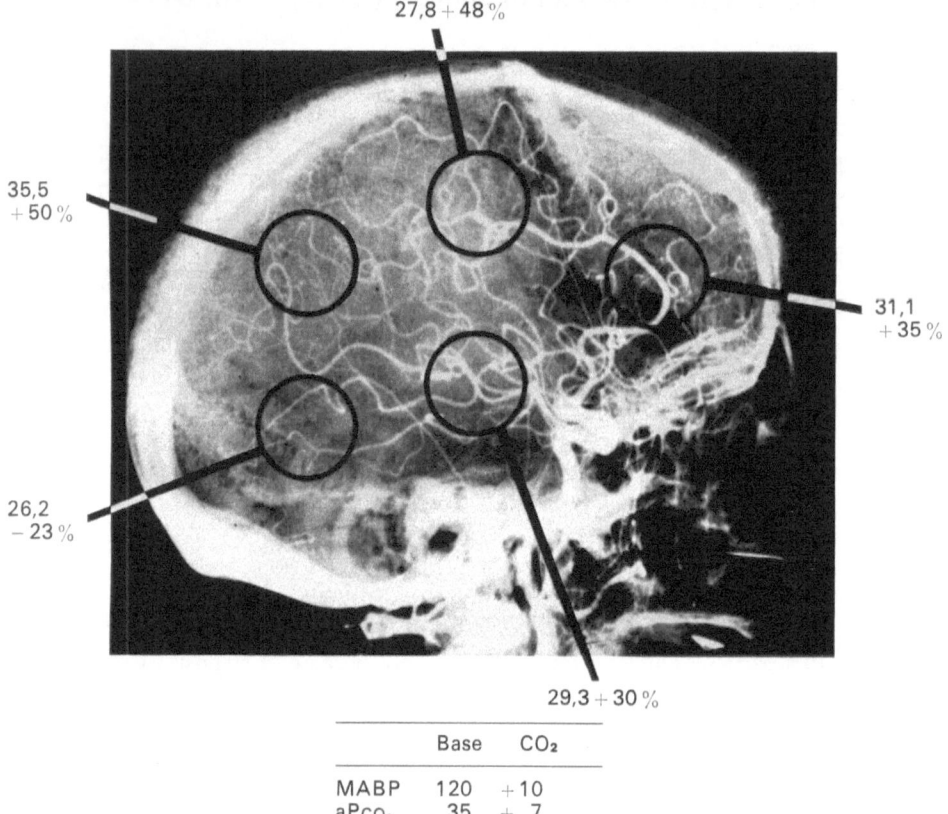

	Base	CO_2
MABP	120	$+10$
aP_{CO_2}	35	$+\ 7$

Fig. 1. Intracerebral steal induced by CO_2 inhalation. Patient of 57 years, whose rCBF has been measured two days after the acute onset of left hemiparesis with sensory and homonymous field defects. The right carotid angiography reveals diffuse atherosclerosis of cerebral vessels without occlusion. The regional cerebral blood flow measured in 5 brain areas by the 85-Kr clearance method shows a reduced perfusion rate in prerolandic and occipital regions in the resting state, while during 7% CO_2 inhalation (Ap CO_2 + 7 mmHg, MABP + 10 mmHg) the rCBF increases in four explored regions. At the same time, in the occipital area the flow undergoes a further reduction of 23% clearly showing an "intracerebral steal" of blood towards normally reacting areas

blood flow passively follow changes in systemic blood pressure, decreasing when system pressure decreases and v.v. (Agnoli *et al.*, 1968).

Severe ischemia, according to this theory, only results, however, when a special kind of event occurs on such a background.

In most cases, this final event may be a sudden, systemic hypotension or a decrease of intracerebral perfusion pressure because of arteriosclerotic narrowing of afferent vessels. In this latter case, ischemia may appear in areas other than that served by the sclerotic artery.

Other possible mechanisms include an active vasodilatation in an area which leads to redistribution of blood with consequent ischemia in other districts supplied by the

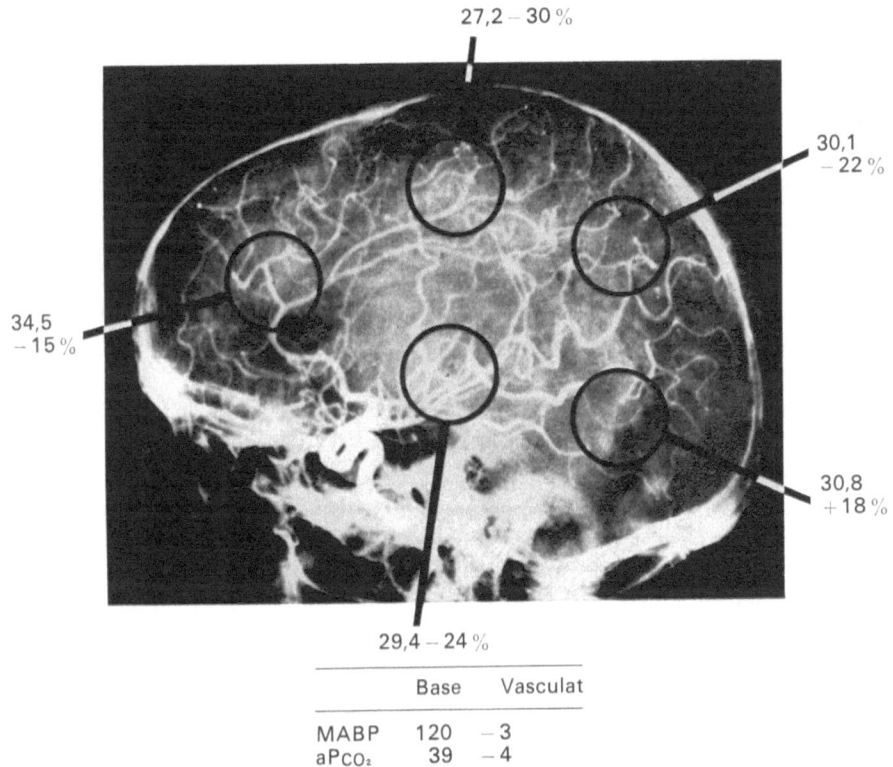

	Base	Vasculat
MABP	120	−3
aPco₂	39	−4

Fig. 2. Counter-steal provoked by hyperventilation. In this patient of 63 years, the left carotid angiography shows relevant signs of arteriosclerosis without arterial occlusion 24 h after the acute onset of right hemiplegia and aphasia. The distribution of blood flow in the six homonymous regions explored is substantially uniform in the resting state: however during hyperventilation induced by i.v. infusion of Vasculat (Ap CO_2 − 4 mmHg; MABP unchanged) the distribution of flow changes remarkably. In fact, while the rCBF diminishes in four of the regions explored, in the occipital region the flow is "paradoxically" increased (+ 18%) showing a "counter-steal" of blood from normally reacting areas

same afferent artery. This is what we properly call "intracerebral steal" and may be reproduced in the laboratory during an active vasodilatation induced by CO_2 inhalation or by papaverine (Fig. 1).

The opposite phenomenon, which we have termed the "counter-steal" may also be noted. It takes place during hyperventilation and after i. v. aminophylline (SKIN-HØJ and PAULSON, 1970); blood flow increases in the area of vasoparalysis, while the nearby regions show the normal vasoconstriction of hypocapnia (Fig. 2). The possibility of regional blood flow decreases in hypercapnia and increases in hypocapnia has important clinical implications, as noted in previous publications (cfr. Fieschi et al., 1969).

Data which can support a haemodynamic theory of cerebral apoplexy must include consideration of the potential importance of an anatomic factor, such as a variation of the circle of Willis, leading to the so-called "enlarged territory" to which we have referred previously. For purposes of analogy, one may examine the phenomenon well known to neurologists and termed "subclavian steal". In this syndrome—due to a unilateral occlusion of the subclavian artery—there is sudden decompensation of flow because of an active vasodilatation due to sudden exertion of the arm homolateral to the artery. In other words, there is an "enlarged" arterial territory, which is dependent solely on one afferent vessel (in this case the opposite vertebral artery). In this territory, a blood steal takes place from one region (the basilar territory) to another (the arterial territory of the arm in exertion). In fact, during rest, the vertebral artery can supply all its new "enlarged" territory, represented by the vertebral-basilar-upper limb circle. However, under conditions of sudden increase in peripheral blood flow requirements (or decrease of systemic flow within the enlarged territory), the most vulnerable areas or those where autoregulation is least efficient will become ischemic.

This concept of enlarged territory was first emphasized in 1964 (Fazio), when the theory of "intracerebral steal" was first introduced to account for certain cerebral ischemic accidents.

In summary, in the field of cerebral apoplexy, complexities of pathogenesis must be taken into account. Cerebral vessels cannot always adapt their diameter to compensate for changes in systemic circulation or in other vascular territories of the brain, and thus focal cerebral ischemia can take place in absence of arterial occlusions.

References

Agnoli, A., Fieschi, C., Bozzao, L., Battistini, N., Prencipe, M.: Autoregulation of cerebral blood flow. Studies in normal subjects and in patients with cerebral vascular disease during drug-induced hypertension. Circulation **38**, 800 (1968).
— Bozzao, L.: L'autoregolazione della circolazione cerebrale. Rec. Prog. Med. 1969.
Brawley, W. B., Straudness, D. E., Kelly, A.: The physiologic response to therapy in experimental cerebral ischemia. Arch. Neurol. Psychiat. (Chic.) **17**, 180 (1967).
Fazio, C.: Observation sur la pathogénèse de l'infarctus cérébral. International Symposium on Questions of cerebral circulation. Salzburg, October 1964.
— Fieschi, C., Agnoli, A.: Discussion sur la pathogénèse des infarctus cérébraux: valorisation des facteurs hémodynamiques. Int. Congr. Neurosurg. Lima 1967.
— — — Bugiani, O., Gottlieb, A.: Fréquence et rôle des anomalies du polygone de Willis et l'artériosclérose dans l'apoplexie cérébrale. Int. Symp. Circ. Cer. Paris 1965, Sandoz Pubbl.
Feindel, W., Perot, P.: Red cerebral veins. A report on arterio-venous shunts in tumours and cerebral scars. J. Neurosurg. **22**, 315 (1965).
Fieschi, C.: Fisiologia della circolazione cerebrale. Proc. Congr. Cardiol., Firenze 1967.
— Agnoli, A., Battistini, N., Bozzao, L., Prencipe, M.: Derangement of the regional cerebral blood flow and of its regulation mechanisms in acute cerebrovascular lesions. Neurology (Minneap.) **18**, 1243 (1968).
— — Prencipe, M., Battistini, N., Bozzao, L., Nardini, M.: Impairment of the regional vasomotor response of cerebral vessels due to hypercarbia in vascular disease. Europ. Neurol. **2**, 13 (1969).

HARPER, A. M.: The interrelationship between PCO_2 and blood pressure in the regulation of flow through the cerebral cortex. Acta neurol. scand. suppl. **14**, 94 (1965).

LASSEN, N. A.: Derangement of the regulation of blood flow in the brain in acute brain disorders. (The luxury perfusion syndrome.) In: Blood flow through the organs and tissues. (HARPER, A. M., BAIN, E. W., Eds.). Livingstone Publ. 1967.

LOEB, C., FAVALE, E.: Contralateral EEG abnormalities in intracranial arterio-venous aneurysm. Arch. Neurol. Psychiat. (Chic.) **7**, 121 (1962).

— MEYER, I. S.: Strokes due to the vertebrobasilar diseases. Co. Thomas Publ. 1965.

MAMO, H., SEYLAZ, J., CARON, J. P., HOUDART, R.: Mesure du débit sanguin du cortex de l'homme par une méthode semi-quantitative continue et focale (Technique, données physiologiques et physio-pathologiques). Rev. neurol. **117**, 689 (1967).

MARSHALL, J.: The management of cerebrovascular disease. Churchill Ltd. 1965.

PASTORINO, P., PIOLA, P.: La persistenza dell'arteria trigeminale primitiva come fattore patogenetico di ischemia cerebrale. Riv. Neurobiol. **12**, 707 (1966).

PAULSON, O. B.: Regional cerebral blood flow studies in various types of cerebrovascular disease. IV Int. Symp. of the Research Group on Cerebral Circulation. Salzburg, Sept. 1968.

SHALIT, N. M., SHIMOJO, S., REINMUTH, O. M.: Carbon dioxide and cerebral circulatory control. Arch. Neurol. Psychiat. (Chic.) **17**, 298 (1967).

SKINHØJ, E., PAULSON, O. B.: The mechanism of action of aminophylline upon cerebral vascular disorders. Acta Neurol. Scand. **46**, 129 (1970).

SYMOND, L.: Experimental study of cerebral arterial spasm. Excerpta med. Int. Congr. Series (Amst.) **139**, 7 (1967).

The Effect of Vascular Perfusion on the Functional and Morphological Sequelae of Transient Cerebral Ischemia*

K.-A. Hossmann** and Y. Olsson***

Transient cerebral ischemia, even of short duration, may be complicated by a serious impairment of the postischemic blood flow ("no-reflow phenomenon") [2, 3, 5, 9]. One important factor in the pathogenesis of this complication is the swelling of the endothelial and the perivascular glial cells during or after ischemia, which may lead to occlusion of the intracerebral capillaries [3, 19]. Since the impairment of the intracerebral circulation after ischemia seems to play a crucial role in determining the maximal duration of ischemia the brain can support [2, 12], an experimental study has been performed designed to prevent this complication [12, 19]. The present communication reports on our results obtained by rinsing the cerebral vessels with various solutions during ischemia.

Material and Methods

Cats were anesthetized with sodium pentobarbital and immobilized by gallamine triethiodide. They were artificially ventilated with room air throughout the experiments. Transient cerebral ischemia was produced by clamping the innominate and the left subclavian artery close to their origin from the aorta. A complete stop of the cerebral circulation was revealed by the lack of pial circulation, the cessation of brain pulsations, a considerable decrease of the brain volume, and by lack of entry into the brain of intravenously injected dyes.

In one group of animals, cerebral vessels were rinsed during ischemia with solutions of different osmolarity at temperature from 27° to 30 °C. In another group of control animals, ischemia was produced in the same way but the perfusion was omitted. The perfusate was introduced into the brain through one carotid artery and was allowed to return into the general circulation. The flow rate of the perfusate was adjusted under visual control of the pial vessels: when the arterial clamping produced complete ischemia, a perfusion pressure was immediately adjusted to yield a flow rate of about 10 ml per min. When ischemia was incomplete, the infusion pressure was increased to exceed the pressure of the remaining blood supply and to

* From the Laboratory of Neuropathology and Neuroanatomical Sciences, National Institute of Neurological Diseases and Stroke, National Institutes of Health, Bethesda, Md. 20014 USA.

** Visiting Scientist from the Max-Planck-Institute for Brain Research, Cologne-Merheim, W. Germany.

*** Visiting Scientist from the Neuropathological Laboratory, Institute of Pathology, University of Gothenburgh, Sweden.

expel all blood elements from the pial vessels. In such instances the flow rate ranged from 15 to 25 ml per min.

The functional impact of the experimental procedure on neuronal activity during and after ischemia was monitored by recording the EEG and the pyramidal response (PR) after electrical stimulation of the sensorimotor cortex; the EEG served as an indicator of spontaneous cellular activity, and the PR provided information about the electrical and synaptical excitability of the neurons which give origin to the pyramidal tract (PT cells).

The electrical stimulation of the motor cortex with electrical pulses of short duration evokes a response in the pyramidal tract which consists of two or more components [20]. The first component (D-wave) is evoked by the electrical excitation of the PT cells. The second and all further components (I-waves) are due to the synaptical (orthodromic) activation of the PT cells by cortical inter-neurons which are activated by the initial electrical pulse. The suppression of the I-waves consequently reflects the abolishment of the synaptical activation of the PT cells, and the suppression of the D-wave reflects the abolishment of their electrical excitability.

At the end of the experiments, the animals were perfused with a cacodylate buffered solution of glutaraldehyde and paraformaldehyde [14], and samples from the frontal cortex, close to the site of the neurophysiological recordings, were taken for light- and electronmicroscopy.

Results

Transient Cerebral Ischemia without Vascular Perfusion

In this study, the time elapsing after the D-wave of the PR was suppressed was used to describe the impact of ischemia on the brain. The suppression of the D-wave reflects the extinction of the electrical excitability of the PT cells: it is a precise measurement of the functional damage inflicted upon these cells, and by using this parameter, we automatically compensate for individual variations which otherwise may influence the ischemic impact, such as incompleteness of ischemia, temperature, depth of anesthesia etc. In this way a possibility is given to compare groups of animals which are uniform with regard to the functional damage. In our experimental situation, the D-wave was suppressed after 6 to 12 min of ischemia, which is considerably later than the changes in the EEG, which became isoelectric after 15 to 25 sec (Fig. 1). The I-wave of the PR was suppressed after 2 to 5 min.

Once the D-wave of the PR was suppressed, ischemia was continued for a pre-determined period ranging from 15 sec to 18 min (total duration of ischemia: 7 to 30 min). The arterial clamps were then released and an immediate recirculation could be observed in the pial vessels. Twenty to sixty min later, however, the pial blood flow gradually slowed down and came to a full stop, although the general circulatory status of the animals did not change.

In most animals in which ischemia exceeded the suppression of the D-wave for less than 10 min (total duration: 7 to 16 min), a partial recovery of this wave occurred. The recovery process started after varying time intervals, but was interrupted and later reversed (Fig. 1). This secondary suppression started as soon as the impairment of the blood flow became apparent in the pial vessels. A recovery of the I-wave or of the EEG was not observed in any of the animals in this group. The morphological changes were in accordance with earlier observations [3, 7, 8, 9]: they consisted in swelling of the endothelial and the perivascular glial cells, and severe changes in the cortical fine structure (Figs. 3 a, 4 a).

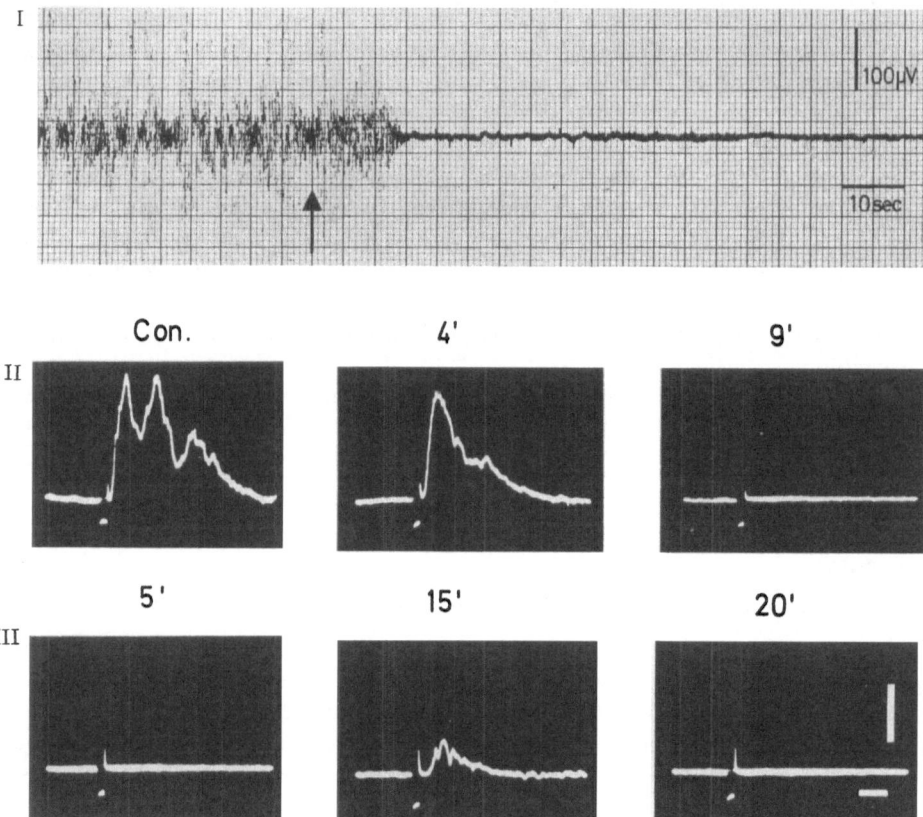

Fig. 1. Transient cerebral ischemia without vascular perfusion. I. Bipolar recording of the EEG from the dura over the right suprasylvian gyrus. The onset of ischemia is marked by an arrow. II. Recording of the pyramidal response (PR) after stimulation of the right soma-tomotor cortex. The control response (Con.) consists of a short latency wave (D-wave) and two I-waves. The amplitude of the I-waves is significantly reduced 4 min after the onset of ischemia and after 9 min all waves are completely suppressed. III. Recording of the PR at different times after restoring the blood flow following ischemia of 15 min duration. A small D-wave has reappeared after 15 min but is suppressed for a second time 5 min later. Calibration for II and III: horizontal bar 2 msec., vertical bar 250 μV

Transient Cerebral Ischemia with Vascular Perfusion

Under the assumption that the swelling of the endothelial and the perivascular glial cells could be counteracted osmotically, the first experiments were carried out using hypertonic solutions. Later, however, isotonic solutions such as saline or Ringer's solution were found to be equally effective. In the perfused animals, the initial impact of ischemia was identical with that in the nonperfused ones. The EEG was suppressed after 15 to 25 sec, the I-wave of the PR after 2 to 7 min and the D-wave of the PR after 6 to 12 min (Fig. 2). Once the D-wave was suppressed, ischemia and vascular perfusion were continued for another period of 7 to 22 min (total duration of ischemia: 14 to 30 min).

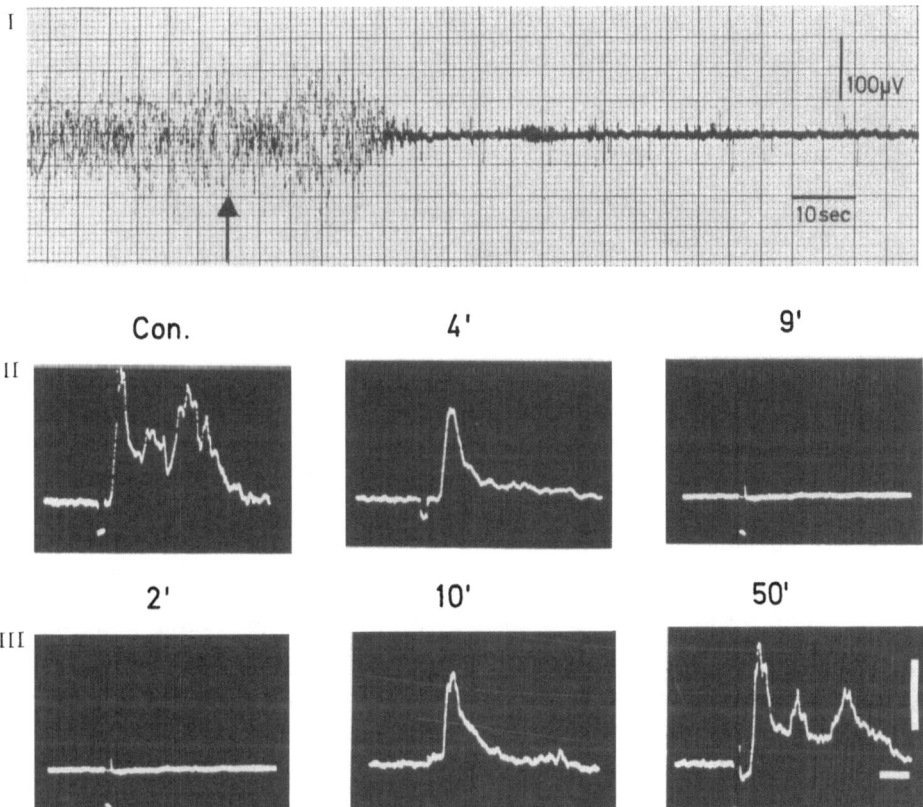

Fig. 2. Transient cerebral ischemia with vascular perfusion of Ringer's solution. I. Recording of the EEG as in Fig. 1. The onset of ischemia and of the perfusion is marked by an arrow. II. Recording of the PR as in Fig. 1. The control response consists of one D-wave and two I-waves. Suppression of the I-waves after 4 min of ischemia and a complete suppression of all waves after 9 min of ischemia. III. Recovery of the PR following ischemia of 25 min duration. The D-wave has reappeared 10 min after restoring the blood flow. The PR completely recovered after 50 min. At this time EEG was also present. Calibration for II and III: horizontal bar 2 msec, vertical bar 200 µV

After restoring the blood flow, the D-wave of the PR recovered within 10 min in all except one animal, even after ischemia of 30 min duration. In several animals with ischemia of 14 to 24 min duration, the I-wave of the PR and even the EEG recovered within $1^{1}/_{2}$ h. Contrary to the first group, the perfused animals did not show a secondary suppression of the recovered signals, although the animals were allowed to survive as long as 3 h after ischemia (Fig. 2).

The pial blood flow remained unimpaired throughout the survival period in all the perfused cats. This fact and several additional observations (distribution of intravenously injected dyes, occurrence of cerebral pulsations, quality of the final fixation by perfusion, etc.) indicated that the intracerebral circulation after ischemia was considerably improved by the saline perfusion.

There was a close correspondence between the morphological findings and the circulatory and neurophysiological data. The intracerebral blood vessels looked normal and there was no abnormal swelling of the perivascular glial cells. Generally,

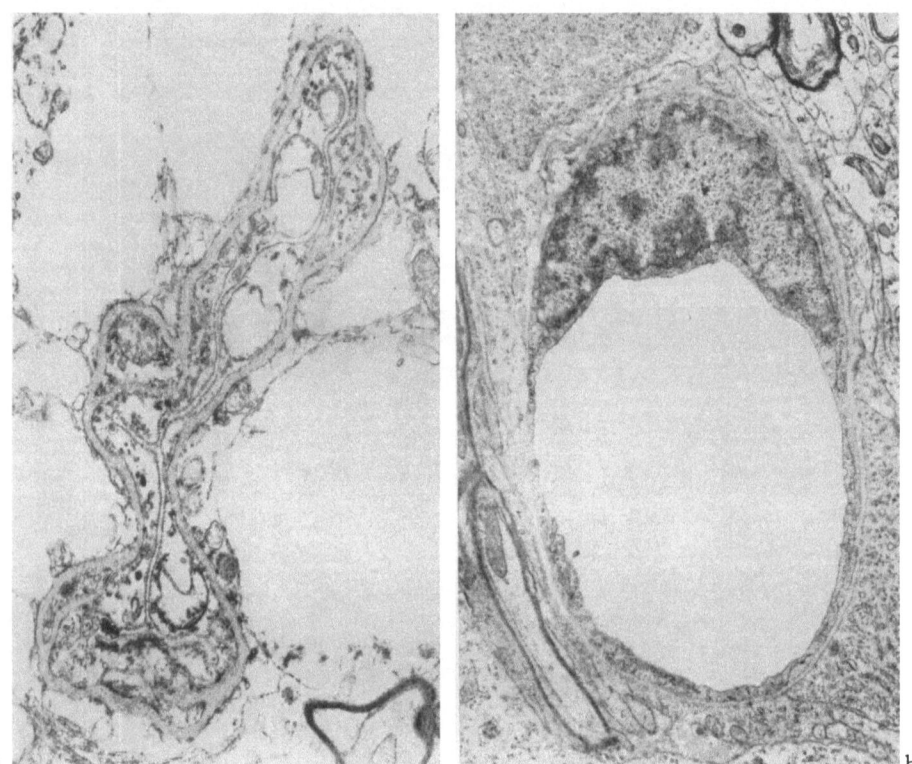

a   b

Fig. 3. Capillaries from the cerebral cortex of cats subjected to transient cerebral ischemia. a) Without vascular perfusion during ischemia. Note the swelling of the perivascular glial cells and of the endothelial cells and the collapse of the vascular lumen. b) With vascular perfusion during ischemia. No ultrastructural changes can be found in the glial and the endothelial cells

the cerebral cortex lacked significant structural alterations even after ischemia as long as 30 min (Figs. 3 b, 4 b).

Discussion

Our observations indicate that an intravascular saline perfusion during ischemia increases the resistance of the brain to this insult, and thus the maximal duration of complete ischemia that the brain can support without suffering irreversible damage can be considerably prolonged. In previous experiments the brain could be revived after total ischemia of only up to 10 min [10], but longer time periods were observed when a slow remaining blood flow was maintained [11]. This was considered to be partly the consequence of a rinsing effect of the perfusate preventing the accumulation of metabolic waste products [21]. Among the various metabolic waste products, lactic acid is probably of particular importance, since it causes a rapid drop of the tissue pH [18, 24]. It has been shown that anaerobic glycolysis with formation of lactic

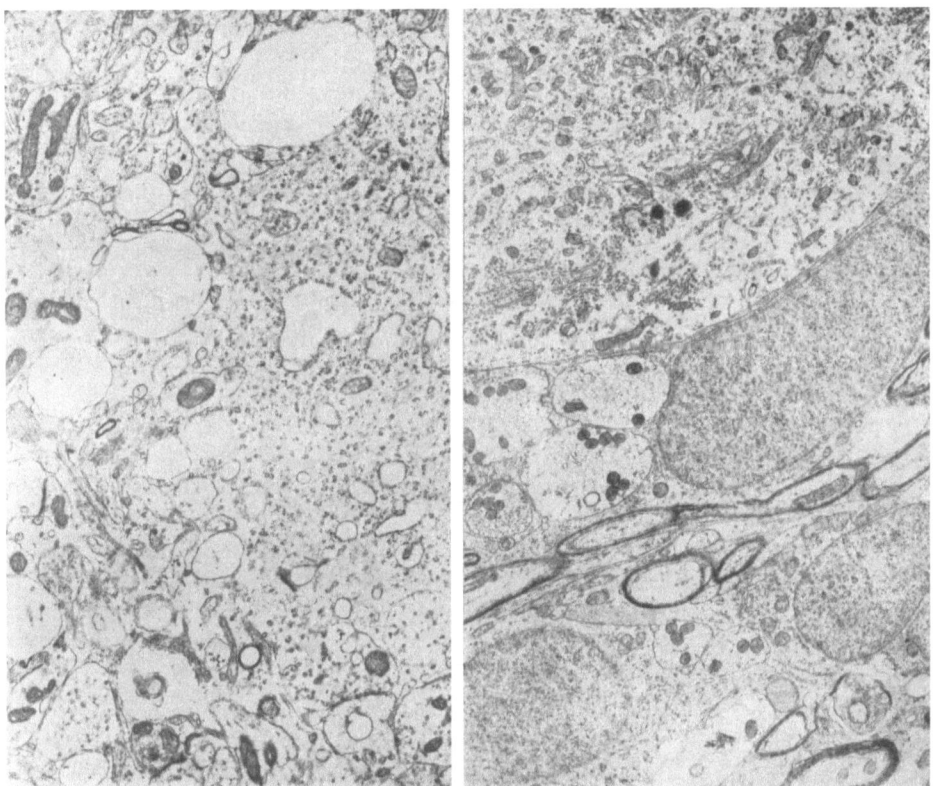

a   b

Fig. 4. Part of the cerebral cortex of cats subjected to transient cerebral ischemia. a) Without vascular perfusion during ischemia. Note the swelling of organelles and the loss of cyto-plasmic ground substance in cellular processes. b) With vascular perfusion during ischemia. Note the absence of fine structural changes

acid may cause severe morphological alterations in the brain [6, 16] and that such changes can be prevented by blocking the anaerobic glycolysis [6].

One most interesting observation in the present study is that ischemia affects the vasculature equally or even earlier than the neurons. Even without vascular perfusion, recovery of the neuronal excitability starts after ischemia up to 16 min duration [12]. However, the recovery process will be interrupted when it coincides with the impair-ment of the postischemic blood flow. The observations in this and previous studies [2, 3] therefore suggest that vascular changes may play a crucial role in determining the maximal duration of ischemia the brain can support. This might have certain clinical implications since it seems that the vascular changes can be influenced by therapeutic maneuvers [13, 19].

The present observations may also provide a clue to the understanding of some observations in which histological signs of ischemic brain damage were less pronounced when ischemia was preceded by a long period of anoxic anoxia [17]. In such cases, the substrate for anaerobic glycolysis may have been exhausted prior to the onset of ischemic anoxia, and the injurious effects of accumulated metabolic waste products may have been eliminated.

References

1. Ames III, A., Gurian, B. S.: Effects of glucose and oxygen deprivation on function of isolated mammalian retina. J. Neurophysiol. **26**, 617—634 (1963).
2. — Wright, R. L., Kowada, M., Thurston, J. M., Majno, G.: Cerebral ischemia. II. The no-reflow phenomenon. Amer. J. Pathol. **52**, 437—453 (1968).
3. Chiang, J., Kowada, M., Ames III, A., Wright, R. L., Majno, G.: Cerebral ischemia. III. Vascular changes. Amer. J. Pathol. **52**, 455—476 (1968).
4. Crowell, J. W., Sharpe, G. P., Lambright, R. L., Read, W. L.: Mechanism of death after resuscitation following acute circulatory failure. Surgery **38**, 696—702 (1955).
5. — Smith, E. E.: Effect of fibrinolytic activation on survival and cerebral damage following periods of circulatory arrest. Amer. J. Physiol. **186**, 283—285 (1956).
6. Friede, R. L., van Houten, W. H.: Relations between post-mortem alterations and glycolytic metabolism in the brain. Exp. Neurol. **4**, 197—204 (1961).
7. Hager, H., Hirschberger, W., Scholz, W.: Electronmicroscopic changes in brain tissue of syrian hamsters following acute hypoxia. Aerospace Med. **31**, 379—387 (1960).
8. Hills, C. P.: (1) Ultrastructural changes in the capillary bed of the rat cerebral cortex in anonix-ischemic brain lesions. Amer. J. Pathol. **44**, 531—543 (1964).
9. — (2) The ultrastructure of anoxic-ischemic lesions in the cerebral cortex of the adult rat brain. Guy's Hosp. Rep. **113**, 333—348 (1964).
10. Hirsch, H., Euler, K. H., Schneider, M.: Über die Erholung und Wiederbelebung des Gehirns nach Ischämie bei Normothermie. Pflügers Arch. ges. Physiol. **265**, 281—313 (1957).
11. — Koch, D., Krenkel, W., Schneider, M.: Die Erholungslatenz des Warmblütergehirns bei Ischämie und die Bedeutung des Restkreislaufs. Pflügers Arch. ges. Physiol. **261**, 392—401 (1955).
12. Hossmann, K.-A., Olsson, Y.: Suppression and recovery of neuronal function in transient cerebral ischemia. Brain Res. In press.
13. — Sato, K.: Recovery of neuronal function after prolonged cerebral ischemia. Science **168**, 375—376 (1970).
14. Karnovsky, M. J.: A formaldehyde-glutaraldehyde fixative of high osmolarity for use in electronmicroscopy. J. Cell Biol. **27**, 137 A (1965).
15. Kowada, M., Ames III, A., Majno, G., Wright, R. L.: Cerebral ischemia. I. An improved experimental method for study, cardiovascular effects, and demonstration of an early vascular lesion in the rabbit. J. Neurosurg. **28**, 150—157 (1968).
16. Lindenberg, R.: Morphostatic necrobiosis. Investigations on the nerve cells of the brain. Amer. J. Path. **32**, 1147—1177 (1956).
17. — Patterns of CNS vulnerability in acute hypoxemia including anesthesia accidents. In: Selective vulnerability of the brain in hypoxaemia, pp. 189—206 (Schadé, J. P., McMenemy, W. H., Eds.). Oxford: Blackwell Scientific Publications 1963.
18. McGinty, D. A., Gesell, R.: On the chemical regulation of respiration. Amer. J. Physiol. **75**, 70—83 (1925).
19. Olsson, Y., Hossmann, K.-A.: The effect of intra vascular saline perfusion on the sequela of transient cerebral ischemia. Acta neuropath. In press.
20. Patton, H. D., Amassian, V. E.: Single and multiple unit analysis of cortical stage of pyramidal tract activation. J. Neurophysiol. **17**, 345—363 (1954).
21. Schneider, M.: Critical blood pressure in the cerebral circulation. In: Selective vulnerability of the brain in hypoxaemia, pp. 1—10 (Schadé, J. P., McMenemy, W. H., Eds.). Oxford: Blackwell Scientific Publications 1963.

22. Speckmann, E.-J., Caspers, H.: Les modifications du potentiel continu cortical pendant l'arrêt respiratoire. Rev. neurol. **117**, 5—19 (1967).
23. Swanson, P. D.: The effects of oxygen deprivation on electrically stimulated cerebral cortical slices. J. Neurochem. **16**, 35—45 (1969).
24. Thorn, W., Heitmann, R.: pH der Gehirnrinde vom Kaninchen in situ während perakuter totaler Ischämie, reiner Anoxie und in der Erholung. Pflügers Arch. ges. Physiol. **258**, 501—510 (1954).
25. Webster, H. de F., Ames III, A.: Reversible and irreversible changes in the fine structure of nervous tissue during oxygen and glucose deprivation. J. Cell Biol. **26**, 885—910 (1965)

Studies of Rapid Changes in Cerebral Circulation and Metabolism during Arousal and Rapid Eye Movement Sleep in Human Subjects with Cerebrovascular Disease*

J. S. Meyer and M. Toyoda

Since the Fick principle was introduced for the quantitative measurement of average cerebral blood flow by Kety and Schmidt in 1945 [1], the nitrous oxide method has provided much information concerning cerebral circulation and metabolism in health and disease [2, 3]. In addition, methods developed by Lassen and co-workers for measuring regional cerebral blood flow using radioactive inert gases have also contributed to knowledge of regional cerebral hemodynamics [4—6]. However, these methods require a steady state of at least 10 min for each measurement so that rapid changes in cerebral blood flow and metabolism cannot be detected.

Recently, methods for measuring rapid changes in blood gases and pH in man have been developed in this laboratory. These methods permit continuous and simultaneous recording of arterial and cerebral venous oxygen tension, oxygen saturation, carbon dioxide and pH. These variables can be compared with the electroencephalogram, arterial blood pressure, electrocardiogram, expiratory carbon dioxide tension and respiration. The validity of this method has already been established in over 200 human volunteers.

This communication will be divided into two parts. First, the methods will be briefly described. Details for the methods have been reported elsewhere [7]. Second, rapid changes in cerebral circulation and metabolism during arousal and rapid eye movement sleep will be reported which serve to illustrate how such transient alterations occur in human subjects. These incidental observations were made during about 450 h of recording cerebral blood flow and metabolism while making therapeutic trials in subjects with occlusive cerebrovascular disease. Since it is unlikely that the opportunity for such observations in man will frequently be available, they are considered to be of sufficient interest to report here.

---

* This work was performed in the Department of Neurology, Wayne State University School of Medicine and the Wayne Center for Cerebrovascular Research, Harper Hospital, Detroit, Michigan, USA.

Dr. Meyer's present adress: Department of Neurology, Baylor College of Medicine, Houston, Texas.

Dr. Toyoda's present adress: Department of Internal Medicine, Keio University, Tokyo, Japan.

Methods

One-half hour prior to study, 50 to 100 mg of meperidine hydrochloride (Demerol) and 0.4 mg of atropine sulfate were given intramuscularly to each subject. Under local anesthesia, using 1 or 2% procaine hydrochloride, the internal jugular vein was catheterized percutaneously using a wire catheter guide and the Seldinger technique. The brachial or femoral artery and the median cubital veins were also catheterized. After the catheter in the jugular vein was inserted above the jugular bulb, 100 mg

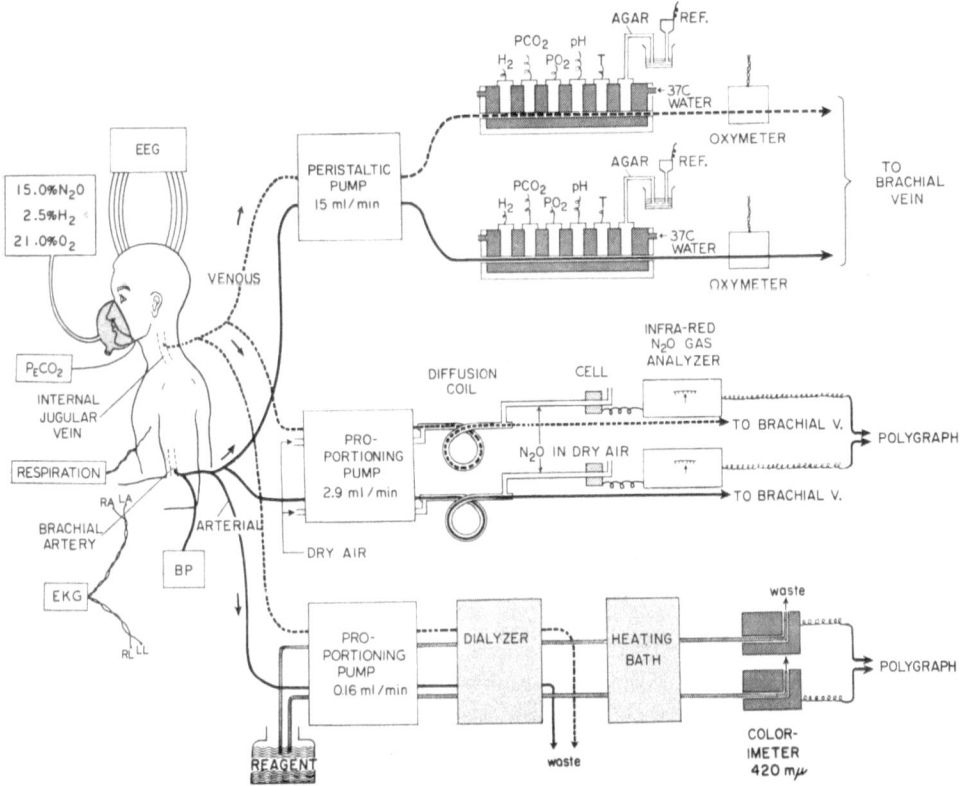

Fig. 1. Schematic drawing of the apparatus for measuring cerebral blood flow and metabolism in human subjects. Reproduced from MEYER, J. S. *et al.*, 1967

of heparin sodium was injected intravenously. At the end of the study, 100 mg of protamine sulfate was injected intravenously to terminate the anticoagulant effect of heparin. The cerebral venous and arterial blood were propelled continuously at 12 to 15 ml/min by means of a peristaltic pump from each catheter, through separate cuvettes for measuring blood gases and pH (Fig. 1). The flowing blood was brought into contact with the sensitive surfaces of the electrodes for measuring oxygen tension, carbon dioxide tension and pH mounted in two separate cuvettes whose temperature was maintained at 37 °C. The blood from both cuvettes was also pumped through oxymeters for measuring oxygen saturation and then returned to the systemic venous circulation without blood loss. Arterial blood pressure was monitored with a strain gauge. Respiration was recorded from changes in transthoracic impedance by means of a transducer. Expired carbon dioxide tension was recorded with an infrared gas analyzer (Beckman Spinco Model LB-1). The electroencephalogram was recorded from 8 standard scalp electrodes by means of an 8-channel electroencephalograph and electrocardiogram was monitored with an electrocardioscope (Electrodyne C-100 M).

Oxygen tension (PO_2) of the blood was measured with an improved polarographic electrode which permits accurate and stable recording. The output was amplified by a Beckman Physiological Gas Analyzer (Model 160) and recorded on a polygraph. Carbon dioxide tension (PCO_2) was measured with a modified Stow-type electrode which has been improved in response time and stability. A small pH electrode (Beckman 39042 Glass Electrode) was used for pH measurement. The reference electrode for pH was connected to the flowing blood by means of an agar salt bridge which was covered by a cellophane membrane and placed in the cuvette. The outputs of PCO_2 and pH electrodes were recorded on the polygraph by pH meters (Radiometer Model 22). Oxygen saturation was measured by a reflection oximeter (Kipp CC Oxymeter MO 3) and the output was recorded on the polygraph by a photogalvanic pickup.

All electrodes were routinely calibrated before and after each study. The entire extracorporeal circulatory system was sterilized by both exposure to ethylene oxide (Ben Venue Sterilizer) and circulatory benzalkonium chloride (Zephiran). Cerebral arterio-venous differences for glucose, lactate and pyruvate were also measured using the Technicon apparatus [8—10] but these measurements are not relevant here.

Cerebral Circulation and Metabolism during Arousal

Typical changes accompanying arousal from sleep in cerebral venous and arterial oxygen tension ($CVPO_2$, PaO_2), oxygen saturation ($CVSO_2$, SaO_2), carbon dioxide tension ($CVPCO_2$, $PaCO_2$) and pH, as well as in the electroencephalogram (EEG), arterial blood pressure and respiration are illustrated in Fig. 2. The subject appeared clinically and from her EEG to be asleep. She was aroused from sleep by talking to her and she then fell asleep again spontaneously. As she awoke, the EEG changed from irregular, mixed theta and alpha activity (A) to dominant alpha activity (B). During EEG activation, $CVPCO_2$ increased and $CVPO_2$, $CVSO_2$ and CVpH decreased. The arterial blood gases and pH showed minimal and totally unrelated changes which reflected the respiratory increase. Blood pressure slightly increased and respiration increased in depth and rate during arousal.

In a previous publication [11], we reported the maximal changes occurring within 3 min of arousal that were determined in 4 subjects with cerebrovascular disease. $CVPO_2$ and CVpH significantly decreased from 35.3 ± 5.5 to 28.4 ± 6.4 mmHg and from 7.319 ± 0.25 to 7.302 ± 0.026, respectively. $CVPCO_2$ also significantly increased from 46.5 ± 7.3 to 48.2 ± 7.8 mmHg. As to the time course of these metabolic changes in cerebral venous blood, the greatest changes usually occurred within 60 sec after the onset of arousal reaction.

It is widely accepted that the internal jugular venous blood derived from the superior bulb is almost free from contamination of extracerebral sources [12] and changes in this blood represent cerebral metabolism. For example, $CVPO_2$ will depend on a) PaO_2, SaO_2 and Bohr effect, b) cerebral oxygen consumption, and c) cerebral blood flow. Hence, during arousal, the decreases of $CVPO_2$ and CVpH associated with the increase of CVPCO$_2$ may be attributed to increased cerebral metabolism, since the changes in arterial blood gases and pH were negligible compared to changes in the cerebral venous blood.

INGVAR and SIESJÖ [13] reported cortical changes during arousal induced by electrical stimulation of the brain-stem in cats; there was a biphasic reaction of the

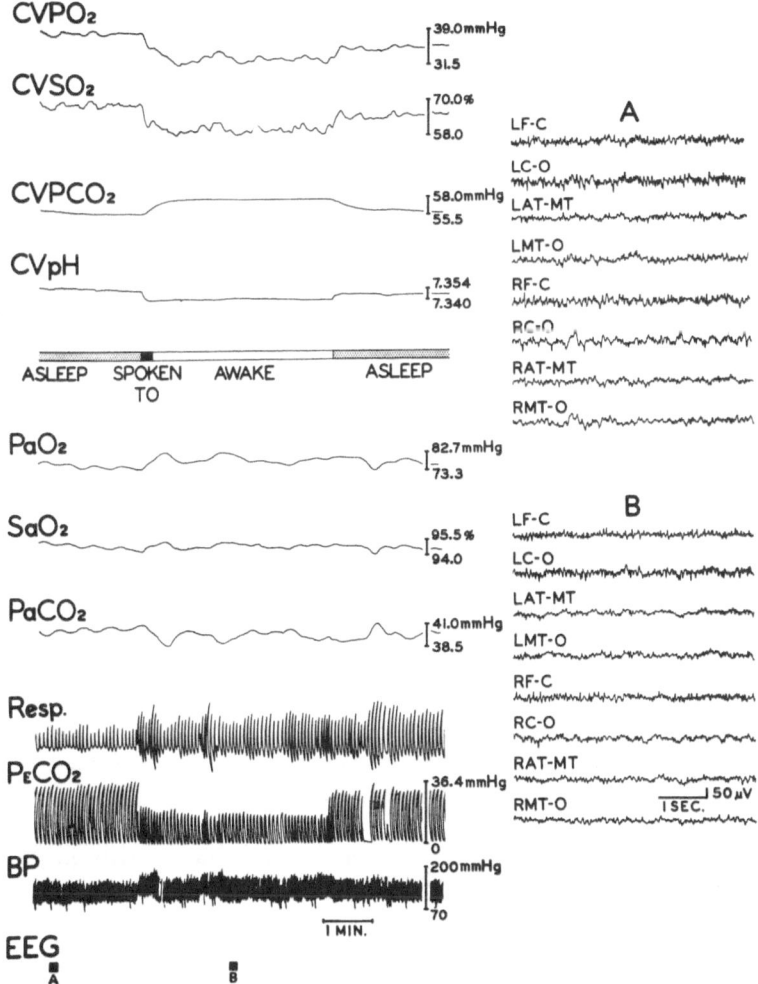

Fig. 2. Continuous records during spontaneous sleep, during arousal, after being spoken to and during resumption of sleep in a 50 year old woman with bilateral stenosis of the internal carotid arteries. Key to this figure and Fig. 3: CVPO$_2$ = cerebral venous oxygen tension, CVSO$_2$ = cerebral venous oxygen saturation, CVPCO$_2$ = cerebral venous carbon dioxide tension, CVpH = cerebral venous pH, PaO$_2$ = arterial oxygen tension, SaO$_2$ = arterial oxygen saturation, PaCO$_2$ = arterial carbon dioxide tension; Resp. = Respiration, PECO$_2$ = carbon dioxide partial pressure of expired air; BP = Arterial Blood Pressure. The EEG records shown on the right (A and B) were sampled at points at the bottom of the polygraphic record. CVPO$_2$, CVSO$_2$ and CVpH decreased while CVPCO$_2$ increased during arousal and vice versa. Note that the small changes in arterial blood were in the opposite direction

cortical PO_2 with an initial decreasing phase, followed by a secondary increasing phase, as the cortical PCO_2 increased. Meyer and Gotoh [14] observed similar cortical changes after an arousal reaction elicited by a noise in the drowsy animal, followed by increased cortical blood flow measured by a thermistor. Arousal and electrical stimulation of the brain stem produced increase in cortical blood flow as indicated by a cerebral venous outflow counter [15, 16]. Recently, it has been shown that there is a significant increase in both cerebral oxygen consumption and blood flow during electrical stimulation of the reticular formation [17]. Although no correlation was detected between changes in cerebral blood flow and oxygen consumption measured by the nitrous oxide method and cerebral functional activity such as mental arithmetic in man [18], a small increase in regional cerebral blood flow using the intra-arterial Xenon[133] clearance technique has been reported during the performance of a digit-span-backward test [19].

As seen in Fig. 2, decreases of $CVPO_2$ and CVpH with an increase of $CVPCO_2$ accompanied the increased neuronal activity indicated in the EEG during arousal and were followed by a return to the steady state with increases of $CVPO_2$ and CVpH as well as decrease of $CVPCO_2$ when sleep was resumed. The later changes are compatible with the increased cerebral blood flow resulting from increased CO_2 and acid metabolites in the brain, and would correlate well with the measured increase of cerebral blood flow reported in animals during arousal.

It was long ago, in 1890, that Roy and Sherrington [20] first adduced changes in cerebral metabolism were followed by concomitant changes in cerebral circulation. This intrinsic mechanism appears to be important in maintaining cerebral circulatory homeostasis in the brain. Accumulation of CO_2 and acid metabolites in the brain increase cerebral blood flow and oxygen delivery and enhance removal of CO_2 and acid metabolites.

Cerebral Circulation and Metabolism during Rapid Eye Movement (REM) Sleep

Recently, we have obtained continuous polygraphic records of changes in cerebral venous blood gases and pH during REM sleep for the first time in man. The subject was a 68 year old male who suffered from a small infarction in the right hemisphere with left facial paralysis. Irregular rapid eye movements of short duration were seen intermittently which were accompanied by low voltage fast activity in the EEG. The conditions observed in this subject were thought to correspond exactly with those described in REM sleep.

There was a remarkable difference from the cerebral circulatory and metabolic changes observed during the arousal reaction. Rapid increases of $CVPO_2$, $CVSO_2$ and CVpH were associated with a decrease of $CVPCO_2$. These changes, which were believed to indicate a rapid and transient increase in cerebral blood flow without metabolic change, reached a maximum within about 20 sec, followed by decreases of $CVPO_2$ and CVpH and an increase of $CVPCO_2$. The arterial blood gases, arterial blood pressure and respiration remained unchanged during the episodes of rapid eye movements (Fig. 3). Furthermore, a concurrent and continuous recording of the cerebral arterial and venous saturation curves for hydrogen gas after inhalation [21] showed a marked decrease in the arteriovenous hydrogen difference due to a rapid rise in the venous hydrogen partial pressure during REM sleep, despite no comparable temporary change in the arterial saturation curve. These changes in cerebral venous blood hydrogen saturation would also indicate temporary increased cerebral blood flow during REM sleep. After completing these observations, the subject was aroused from the drowsy state by a loud noise and then the typical changes obtained during arousal reaction were observed.

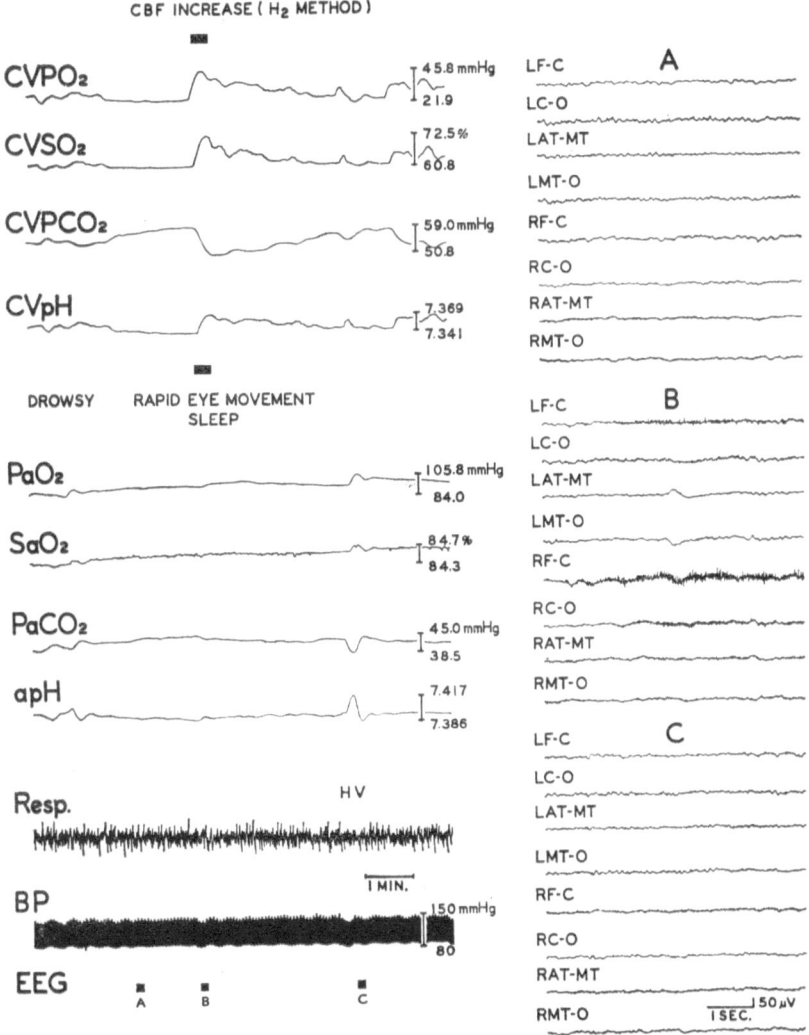

Fig. 3. Continuous records during spontaneous rapid eye movement sleep in a 68 year old male with left facial paralysis due to a small area of infarction of the right cerebral hemisphere. The key to the figure is the same as Fig. 2. The rapid eye movements consisted of rapid oscillation of the eye, visible to two observers, since the left eye was open. Note the desynchronization of the EEG at B during rapid eye movement sleep. CBF was being measured by the hydrogen inhalation method when REM sleep began; a brief narrowing of the cerebral arteriovenous difference for hydrogen occurred for about 30 sec during REM sleep. Toward the end of the record, there was a brief interval of hyperventilation marked HV on the record. The changes indicate a considerable but transient increase in CBF during REM sleep

11   Cerebral Circulation and Stroke

Since the first report [22] that visual dreams were accompanied by low voltage fast activity in EEG and rapid eye movements, some information has been gained concerning cerebral neuronal activity, blood flow and metabolism during this phenomenon.

There is evidence that the discharge rate of neurons increased markedly during REM sleep, almost approaching that of wakefulness [23]. KANZOW et al. [24] found that cortical blood flow increased during the desynchronized stage of sleep in spite of a fall in arterial blood pressure. A similar increase of the regional blood flow was observed in the rhomboencephalic reticular formation during paradoxical sleep [25]. Furthermore, KAWAMURA and SAWYER [26] reported greater elevations in the temperature in various parts of the rabbit brain than those observed during arousal.

REIVICH et al. [27] have recently studied the regional cerebral blood flow in the cat during REM sleep using an autoradiographic technique. There was a significant increase in regional blood flow in 25 regions which mainly occurred in the structures located in the brain stem and diencephalon. Total cerebral blood flow was estimated to increase about 80%. On the other hand, the values for arterial PCO_2, PO_2 and pH during REM sleep were not different from those in the controls.

In our case, the direction and degree of these rapid changes in cerebral blood flow were calculated from cerebral arteriovenous differences for oxygen content. The arteriovenous oxygen content is inversely proportional to cerebral blood flow, if cerebral metabolic rate for oxygen remains constant. These methods have limitations but permit some estimation of rapid change in cerebral blood flow during short intervals [28], and appeared justified since the arteriovenous hydrogen differences also indicated a remarkable increase in blood flow. Using this method, total cerebral blood flow increased by approximately 50% during the interval of 20 sec after the onset of rapid eye movement, which seems to be in good agreement with the findings reported by REIVICH et al.

An increase in cerebral blood flow of such magnitude has been observed only during convulsions induced by drugs and electroshock [9, 29, 30] or inhalation of carbon dioxide gas mixtures [31]. Since hypercapnia and convulsions could be excluded, the marked increase in cerebral blood flow during REM sleep is possibly due to neurogenic cerebral vasodilatation. If it were due to increased neuronal metabolism, it should be preceded by an increase in cerebral venous PCO_2 which occurred in arousal from sleep but not in rapid eye movement sleep.

References

1. KETY, S. S., SCHMIDT, C. F.: The determination of cerebral blood flow in many by the use of nitrous oxide in low concentrations. Amer. J. Physiol. 143, 53—66 (1945).
2. LASSEN, N. A.: Cerebral blood flow and oxygen consumption in man. Physiol. Rev. 39, 183—238 (1959).
3. SOKOLOFF, L.: The action of drugs on the cerebral circulation. Pharmacol. Rev. 11, 1—85 (1959).
4. LASSEN, N. A., INGVAR, D. H.: The blood flow of the cerebral cortex determined by radioactive krypton[85]. Experientia (Basel) 17, 42—43 (1961).
5. INGVAR, D. H., LASSEN, N. A.: Regional blood flow of the cerebral cortex determined by krypton[85]. Acta physiol. scand. 54, 325—338 (1962).
6. LASSEN, N. A., HØEDT-RASMUSSEN, K., SØRENSEN, S. C., SKINHØJ, E., CRONQUIST, S., BODFORSS, B., INGVAR, D. H.: Regional cerebral blood flow in man determined by krypton[85]. Neurology (Minneap.) 13, 719—727 (1963).

7. GOTOH, F., MEYER, J. S., EBIHARA, S.: Continuous recording of human cerebral blood flow and metabolism: Methods for electronic monitoring of arterial and venous gases and electrolytes. Med. Res. Engineering 5, 13—19 (1966).
8. MEYER, J. S., GOTOH, F., AKIYAMA, M., YOSHITAKE, S.: Monitoring cerebral blood flow, oxygen and glucose metabolism. Analysis of cerebral metabolic disorder in stroke and some therapeutic trials in human volunteers. Circulation 36, 197—211 (1967).
9. — — — — Monitoring cerebral blood flow, oxygen, glucose, lactate and ammonia metabolism. Experimental trials in animals. Circulat. Res. 21, 649—660 (1967).
10. — SAWADA, T., KITAMURA, A., TOYODA, M.: Cerebral oxygen, glucose, lactate and pyruvate metabolism in stroke. Therapeutic considerations. Circulation 37, 1036—1048 (1968).
11. — GOTOH, F., TOMITA, M.: Cerebral metabolism during arousal and mental activity in stroke patients. J. Amer. Geriat. Soc. 14, 986—1012 (1966).
12. SHENKIN, H. A., HARMEL, M. H., KETY, S. S.: Dynamic anatomy of cerebral circulation. Arch. Neurol. Psychiat. (Chic.) 60, 240—252 (1948).
13. INGVAR, D. H., SIESJÖ, B.: Measurements of oxygen and carbon dioxide tension on the pial surface. EEG clin. Neurophysiol. 11, 837—838 (1959).
14. MEYER, J. S., GOTOH, F.: Interaction of cerebral hemodynamics and metabolism. Neurology (Minneap.) 11 (part 2), 46—65 (1961).
15. INGVAR, D. H., SÖDERBERG, U.: A new method for measuring cerebral blood flow in relation to the electroencephalogram. EEG clin. Neurophysiol. 8, 403—412 (1956).
16. — — Cortical blood flow related to EEG pattern evoked by stimulation of the brain stem. Acta physiol. scand. 42, 130—143 (1958).
17. MEYER, J. S., NOMURA, F., SAKAMOTO, K., KONDO, K.: Effect of stimulation of the brain-stem reticular formation on cerebral blood flow and oxygen consumption. EEG clin. Neurophysiol. 26, 125—132 (1969).
18. SOKOLOFF, L., MANGOLD, R., WECHSLER, R. L., KENNEDY, C., KETY, S. S.: The effect of mental arithmetic on cerebral circulation and metabolism. J. clin. Invest. 34, 1101—1108 (1955).
19. INGVAR, D. H., RISBERG, J.: Increase of regional cerebral blood flow during mental effort in normals and in patients with focal brain disorders. Exp. Brain Res. 3, 195—211 (1967).
20. ROY, C. S., SHERRINGTON, C. S.: On the regulation of the blood supply of the brain. J. Physiol. (Lond.) 11, 85—108 (1890).
21. GOTOH, F., MEYER, J. S., TOMITA, M.: Hydrogen method for determining cerebral blood flow in man. Arch. Neurol. (Chic.) 15, 549—559 (1966).
22. ASERINSKY, E., KLEITMAN, N.: Two types of ocular motility occurring in sleep. J. appl. Physiol. 8, 1—10 (1955).
23. EVARTS, E. V.: Activity of neurons in visual cortex of cat during sleep with low voltage fast activity. J. Neurophysiol. 25, 812—816 (1962).
24. KANZOW, E., KRAUSE, D., KÜHNEL, H.: Die Vasomotorik der Hirnrinde in den Phasen desynchronisierter EEG-Aktivität im natürlichen Schlaf der Katze. Pflügers Arch. ges. Physiol. 274 593—607 (1962).
25. BAUST, W.: Local blood flow in different regions of the brain-stem during natural sleep and arousal. EEG clin. Neurophysiol. 22, 365—372 (1967).
26. KAWAMURA, H., SAWYER, C. H.: Elevation in brain temperature during paradoxical sleep. Science 150, 912—913 (1965).
27. REIVICH, M., ISAACS, G., EVARTS, E. V., KETY, S. S.: Regional cerebral blood flow during REM and slow wave sleep. Trans. Amer. neurol. Ass. 92, 70—74 (1967).
28. MEYER, J. S., GOTOH, F., TAKAGI, Y., KAKIMI, R.: Cerebral hemodynamics, blood gases and electrolytes during breath-holding and the Valsalva maneuver. Circulation 38 (Suppl. 2), 35—48 (1966).
29. SCHMIDT, C. F., KETY, S. S., PENNES, H. H.: The gases metabolism of the brain of the monkey. Amer. J. Physiol. 143, 33—52 (1945).
30. PLUM, F., POSNER, J. B., TROY, B.: Cerebral metabolic and circulatory responses to induced convulsions in animals. Arch. Neurol. (Chic.) 18, 1—13 (1968).
31. KETY, S. S., SCHMIDT, C. F.: The effects of altered arterial tensions of carbon dioxide and oxygen in cerebral blood flow and cerebral oxygen consumption of normal young men. J. clin. Invest. 27, 484—492 (1948).

Influence of Aggregated Blood Cells on Cerebrovascular Disease*

L. Swank and W. Bartsch

The purpose of this paper is to examine the hypothesis that adhesive and/or aggregated platelets formed *in vivo* can contribute to cerebral ischemia by occluding the microvasculature of the brain. It is known that intravascular aggregation of blood cells (the "sludging" of Knisely) with slowing of blood flow in the microcirculation is frequently present in vascular disease [1, 2, 3]. It has also been shown that "sludging" of the blood is frequently associated with platelets which become unusually adhesive to the presence in the blood of added adenosine diphosphate (ADP) [4]. Since in cardiovascular disease platelets are often adhesive [5], one is tempted to conclude that the thread common to all phenomena, namely platelet adhesiveness, is causally associated in some way with the ischemias of vascular disease, and in particular cerebrovascular disease. Before jumping to this conclusion it is wise to examine the evidence more closely.

For some years the significance of intravascular blood cell aggregation has been a matter of controversy. This has been due in large part to the following factors. First, the method lacks objectivity, in that what is seen, and its interpretation is subject to wide variations. Second, many observers have attached equal importance to the aggregation seen in arteries and veins. Knisely, Bloch, and Davis and Landau and others have stressed that venous aggregation is of questionable importance, whereas "sludging" on the arterial side of the circulation is of pathophysiological importance. When these two pitfalls are avoided, examination of the microcirculation can be of significance.

Aggregated Blood Cells in Myocardial Disease

Bloch [1] examined the conjunctival microcirculation of 75 patients with acute myocardial infarctions. During the first 24 to 48 h the circulation and vessel walls remained normal. Twenty-four to 48 h later the majority of red cells were aggregated in the arterioles and the circulation was slowed. In many vessels cessation of circulation for up to 30 sec was observed. In the venules the changes were greater, and often the blood appeared like a paste and was deeply cyanotic. These changes were accompanied by increased permeability of the venules and by edema of the surrounding tissues, often so severe in places that optical resolution of the events was very difficult. About two weeks later the changes had started to resolve. Bloch studied 205 controls including those with hypertension, with and without angina

* With technical assistance of Kathleen H. Prichard and George Whitney from Division of Neurology, University of Oregon Medical School, Portland, Oregon. Supported by PHS grant H-6336.

pectoris, and others with valvular heart disease, in which the changes just described did not occur. He stressed the point that the blood changes were the result of the myocardial infarction, not its cause. Also, he noted that the intravascular blood cell aggregation contributed to the seriousness of the condition by increasing the peripheral resistance to the flow of blood. Other investigators, including KNISELY, DAVIS and LANDAU, have made similar observations in patients with acute myocardial infarctions.

Relationship of Aggregated Blood Cells to Adhesive Platelets

Intravascular aggregation of blood cells, such as that just described, was shown by SWANK and DAVIS [4] to be associated usually with slightly increased (circa 10%) screen filtration pressure (SFP). When small doses of ADP (1 γ/ml) were added to blood, marked increases (300 to 1000%) in the SFP resulted. The SFP is a measure of the resistance to flow, in mmHg, of blood which is forced at a constant speed through a screen with multiple pores 20×20 micra square. This measurement normally indicates primarily the bulk viscosity of the blood. When platelets are adhesive or aggregated, however, the SFP is increased many fold and then reflects primarily these adhesive properties of the blood. The addition of small doses of ADP to blood (1 to 4 γ/ml) usually causes a slight increase in SFP due presumably to a slight increase in platelet adhesiveness. In patients with arterial intravascular aggregation ADP caused far greater increase in adhesiveness of platelets and hence a correspondingly greater increase in the SFP.

In other studies with the use of platelet rich plasma and the turbidity method of measuring platelet aggregation, McDONALD and EDGILL [5] have shown that acute myocardial infarction is associated with an increased adhesiveness of platelets for glass beads, and HAMPTON and MITCHELL [6] have found an increased sensitivity of platelets to added ADP as demonstrated by changes in their electrokinetic mobility. These observations would appear to link intravascular blood cell aggregation with abnormal adhesiveness of platelets, and these in turn with myocardial infarction. However, these events may result from, rather than cause the myocardial infarction.

Adhesive Platelets in "Shock" and Ischemia

It would be interesting to look into other conditions in which platelets become abnormally adhesive and/or sensitive to ADP. SWANK, HISSEN and BERGENTZ [7] observed a gradual and progressive increase in the SFP of both venous and arterial blood following severe soft tissue trauma to the lower extremities of dogs. More dramatic, however, was the fact that the SFP, after additions of 1 to 2 γ ADP/ml to heparinized blood *in vitro*, progressively increased until shortly before death of the animals. This potentiation of the SFP by ADP was comparable in degree to that observed in humans with intravascular blood cell aggregation by SWANK and DAVIS [4]. In other studies SWANK and his collaborators [8, 9] noted a marked increase in the SFP of blood during acute hypotension which was also attented by potentiation of the SFP with ADP. HIRSCH and his collaborators studied the effects of total temporary ischemia of an isolated head [10], hind limb, or kidney, and of anoxia [11] on the SFP of blood leaving the ischemic or anoxic organ. In each instance a significant increase in SFP, due in all probability to increased adhesiveness of the platelets, occurred. In studies in which a hind limb of a dog was rendered ischemic by application of a tourniquet for several hours, after release of the tourniquet the blood leaving the leg had a significantly increased SFP, and usually the dogs were found dead in their cages the next morning[1]. In similar experiments Goodman et al. [12] demonstrated hemorrhages in the lungs and platelet aggregate emboli in the lung microvasculature by electron microscopy. These last observations would explain the increase in pulmonary artery pressure and simultaneous decrease in left atrial pressure observed by Hissen and Swank [13] during surgical "shock" from severe trauma, or after acute hypotension, and the increased artery-alveolar P_{CO_2} difference observed by Swank and Edwards [14] in dogs after a short period of severe acute hypotension.

Thus we have a group of situations, in all of which the circulation is seriously compromised either in the entire animal, or in one organ or area of the animal. In each instance the platelets become adhesive and aggregate, and occlusion of the microvasculature of the lungs occurs, presumably by embolization. If the number of emboli is not great the lungs are able to filter and remove the emboli from the circulation. *However, evidence is available from SFP studies that these emboli can slowly work their way through the vascular system of the lung and then appear in arterial blood where they are free to embolize other organs of the body including the brain and spinal cord* [15, 13].

Intravascular Aggregation after Trauma

In most conditions described above in which platelets were abnormally adhesive, intravascular aggregation of blood cells has also been noted. This has been particulary marked after trauma, in which instance it has been described in considerable detail by Knisely, Eliot, and Bloch [2]. After traumatizing the omentum they observed the immediate formation of a straw colored adhesive envelope around the red cells passing intravascularly through this traumatized area. The red cells clumped together and the circulation slowed, and in some vessels stopped. No unusual surrounding tissue edema was observed. From the damaged blood vessels clumps of red cells embedded in "sludge" were continuously fed into the venous drainage of the area. These observations, plus others to be described later in connection with circulatory changes after fat meals, raises the question of whether the microembolization of the lungs might not be due to aggregated red cells *per se*. It also raises the question of the relationship of adhesive and aggregated platelets to aggregated red cells such as those observed in the conjunctival blood vessels.

Mechanism of Aggregation of Red Cells: Relationship to Platelets

It seems unlikely that aggregated red cells alone, without additional intracellular forces or cementing material, would be sufficiently cohesive to cause obstruction of the microcirculation. Treatment of red cells *in vivo* with neuraminidase, which markedly decreases their negative electrokinetic potential and causes very marked aggregation of the red cells, is known to not obviously alter the physiological state of otherwise intact animals [16]. Also when red cells are treated with neuraminidase *in vitro* an approximate 30% increase in the SFP occurs [17]. In isolated organ perfusion studies, this degree of change in the SFP of perfused blood has little effect on the function of an organ [18]. On the other hand, the presence of adhesive platelets in blood, the

---

[1] Unpublished personal observations.

result of traumatic or hypotensive "shock" or marked ischemia of an organ or extremity, markedly increases the SFP of the blood. The circulation is often impaired sufficiently to cause pathophysiological changes both in isolated organs [15], and in the intact animal [13, 14], and may lead to death of the animal. Removal of the adhesive and aggregated platelets, by wool (glass or Dacron) filtration returns both the SFP and the flow characteristics of blood in an isolated organ to normal, and prolongs function[2].

This leads to a consideration of whether the clumps of red cells are held together by platelet "cementing material" (see Fig. 4 in STEHBENS paper [19],) or are preceeded by a "plug" of adhesive and/or aggregated platelets which are the actual cause of the slowed circulation. Such "material" or "plugs" have been infrequently seen, possibly because of inadequacy of the optical equipment presently available for viewing the conjunctival circulation. Even when viewed with transmitted light at much higher magnification, platelets may be seen free in the plasma and rarely are seen attached to the red cells but, according to STEHBENS [19], have a close affinity for leukocytes. However, it is well to keep in mind that platelet aggregates formed *in vivo* and occluding the microvasculature of the lungs have been seen by both light and electron microscopic examination [12].

Slowing of the circulation by a leading aggregate of adhesive platelets could explain the clumping of red cells, increased permeability of the venules, edema of the surrounding tissues, and hemoconcentration by plasma skimming such as BLOCH [1] described in patients after myocardial infarction, and KNISELY et al. [2] described after trauma to blood vessels of the omentum. It seems unlikely that hemoconcentration alone would do more than increase the peripheral vascular resistance and contribute to slowing of the circulation. HELLEM [20] has stated that platelets become more adhesive as the concentration of red cells increases, and marked slowing to cessation of the circulation has been shown to increase the adhesiveness of platelets and their sensitivity to ADP as measured by the SFP technique [4, 7]. Possibly the platelets are thereby activated, and in the presence of a high concentration of large molecular weight proteins, the result of increased permeability of the blood vessels, so called "sludge" is produced. As masses of this "sludge" break loose and enter the circulation, the trend to thrombus formation is reversed, and the "sludge" dissolves. The adhesive and/or aggregated platelets remain unaffected and lodge in the micro-circulation of the lungs.

DITZEL [21] observed an unspecified coating of "protein" surrounding red cells in diabetic subjects with intravascular red cell aggregation in which the protein patterns were markedly altered. This coating consisted of the same proteins, i.e., a 2-globulin and fibrinogen, which were found to be abnormally increased in the plasma. This observation would appear to be consistent with the concept of "sludge" formation described above.

Mechanism of Platelet Adhesiveness

The mechanism of platelet adhesiveness which has been assumed to initiate the events leading to microembolism is also left unanswered. BOLTON et al. [22] has advanced

---

[2] A Dacron wool filter to remove platelet-leukocyte aggregates from blood being transfused is under development with Pioneer Filters, Inc., P.O. Box 456, Hillsboro, Oregon 97123, USA.

the hypothesis that the abnormal sensitivity of platelets from patients with ischemic heart disease to ADP is due to a factor in the plasma composed of two parts. One is low-density lipoprotein containing lecithin. The other component is labile and is present in all bloods after centrifugation. The first factor can be transferred to a normal blood to make its platelets abnormal. One must also consider the release of serotonin from platelets, triggered by ischemia and anoxia from vascular stasis, as an early or initial factor in production of platelet adhesiveness. Serotonin is released to the general circulation in relatively large amounts during acute hypotension and after severe trauma [9, 7], and serotonin is known to cause marked adhesiveness of platelets when tested by the SFP method [23]. This latter observation has been confirmed with platelet rich plasma [24]. Serotonin has a half life adhesive producing effect on platelets in whole blood of approximately 10 min. ADP in comparable dosages has a half life effect nearer to 2 min, and when serotonin is blocked by a serotonin antagonist, the half life of the ADP is nearer to 1 min, and the degree of adhesiveness produced is much less marked [23, 25]. The relationship of these factors to one another is not known, nor is their relationship to human disease established. This general subject is treated in some detail in another paper [26]. Some changes, strikingly similar to those produced by traumatic or hypotensive "shock", occur in dogs when injected with serotonin intravenously [23]. These are hemoconcentration, intravascular blood cell aggregation, hemorrhages in the lung similar to those seen in "shock", and a P_{CO_2} alveolar air-arterial difference[3]. When injected into a carotid artery increased permeability of the cerebrovascular system to trypan blue, probably due to microembolization of the cerebral microcirculation by platelet aggregates, occurs [27]. In addition, the electrokinetic mobility of both red cells and platelets have been shown to decrease by up to 40% during "shock" from trauma [28] and acute hypotension [29], and the addition of ADP to platelet rich plasma has had a similar effect on platelets [30]. These reductions in negative electrical potential on blood cells would tend to increase the contact chance and hence aggregation of platelets and also red cells.

Aggregation of Red Cells after Lipid Meals

There is yet one condition to be considered which is important to our consideration of aggregation of blood cells. After large butter fat meals the circulating blood cells in hamsters [31], dogs [32, 33], rabbits [34], cats [35], and man [36, 37] have been observed to aggregate, and in hamsters, rabbits, cats and man the circulation has been visibly or measurably slowed. In hamsters a viscosity, relative to water, of 15 or more was often measured 6 h after fat meals, and cyanosis and hyperemia of the abdominal viscera was observed. The addition of sodium citrate or heparin to such bloods followed by repeated measurements of viscosity revealed a return of such high relative viscosities to the value of 5 to 7 in 5 min. (Normal relative viscosity in hamsters was approx. 4 times that of water by the capillary tube method [38]). This rapid return of viscosity to near normal upon addition of anticoagulants suggested that the coagulation process had been initiated, but could still be reversed at this point. In rabbits, marked slowing of the circulation was detected using the ear oximeter and dye injection technique [34] after fat meals, but in dogs this method failed to reveal more

---

[3] Unpublished observations in collaboration with Dr. Miles Edwards.

than a tendency for the circulation to slow down after both butter fat and oil meals[4], although after large daily lipid meals deep cyanosis of the skin of the extremities and ears of the dog developed. *In vitro* examination of wet smears of whole blood revealed adhesiveness of the red cells for one another with often an amorphous film formation around them [32]. An amorphous material also coated hematocrit tubes and viscosity needles [33, 34], and could not be removed by water or the usual fat solvents or by detergents, but was readily removed by the proteolytic enzyme, papine. Of possible importance, also, was a marked increase in fragility to mechanical trauma of red cells which occurred during periods when the plasma was lipemic [39]. Since ADP causes

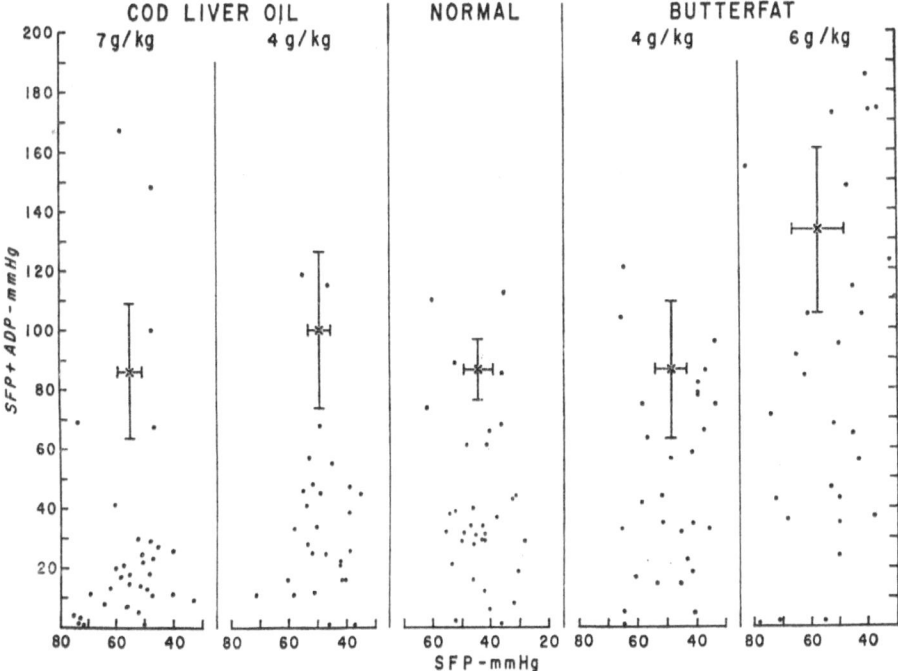

Fig. 1. Shows SFP and SFP + ADP for 5 dogs repeatedly tested before and after lipid meals. For details see Table 1

platelet adhesiveness, this phenomenon could at times be of considerable importance, especially if serotonin were also present.

Other changes suggesting and consistent with significant circulatory slowing were a marked reduction in the available oxygen in the cerebrum of hamsters [40], and convulsions [41], elongation of the QT interval in the electrocardiogram [42], and marked increases in potassium excretion after fat meals [43]. These effects were all thought to be due to the slowing of the circulation and anoxia from intravascular aggregation of blood cells. After oil meals, none or much less marked reductions in the oxygen availability in the brain occurred and other changes noted after butter fat meals were absent. Platelet counts were often reduced by as much as 25 to 50% after

---

[4] Unpublished observations by BARTSCH and SWANK.

fat meals, and increases followed by decreases in the red cell concentrations occurred [32].

Four to 8 h after feeding large fat or oil meals the SFP of the blood of dogs increased from 20 to 40% (Table 1), depending upon the size of the meals. Upon addition of ADP to the blood an increase in SFP, far greater than was observed normally, also occurred after large butter fat meals. Changes after oil meals were absent or less marked (Fig. 1, Table 1). In keeping with our previous remarks this would appear to indicate that a significant increase in the platelet adhesiveness was produced by the butter fat meals, but not by cod liver oil meals. Another condition in which intravascular aggregation was attended by platelets abnormally sensitive to ADP is added to our list.

Table 1. *The mean screen filtration pressure (SFP) and mean SFP + adenosine diphosphate (ADP) with standard deviation (S.D.) and limits of confidence (p values) for 5 dogs before and after feeding cod liver oil and butter fat meals. The technique of ADP addition was the same as used before [4]. Each dog, after each feeding and as a control, was tested 4 to 8 times and their averages were used to compute the figures used in the table. All determinations, however, were plotted in Fig. 1*

	Control	Cod liver oil		Butter fat	
		4 m	7 m	4 m	6 m
SFP	44.1	49.0	55.1	48.4	57.1
S.D.	4.8	3.5	2.9	4.1	8.2
p value		>.05	>.05	>.05	>.05
SFP + ADP	86.1	100.1	86.6	96.2	132.6
S.D.	10.4	27.7	21.5	19.8	27.9
p value		>.05		>.05	>.05

Blood Viscosity in Patients with Cerebral and Other Vascular Disease

Only a few studies of measurable physical chemical changes in blood of patients with cerebrovascular and other vascular diseases have been made, but these have indicated that the gross viscosity of the blood is elevated. Using a capillary tube method [44], SWANK [45] found that plasma viscosity was increased by approximately 10%. When corrections were made for differences in the concentration of red cells (hematocrit), a similar magnitude of change in whole blood viscosity was also present. In these studies no anticoagulant was added to the blood. DINTENFASS [46] used a cone-in-cone viscometer with low rates of shear (0.1 sec^{-1}) to study blood without anticoagulant in patients with vascular disease. He observed a much greater increase in viscosity in his patients, and ascribed this to aggregation of red cells. At higher rates of shear more modest increases in viscosity were noted, which he felt were due to changes in the plasma alone. SWANK [45] placed his patients on a low fat diet and studied their blood periodically for 6 to 18 months, during which time a significant (circa 10%) decrease in plasma and whole blood viscosity occurred. In other studies [47] using the same capillary tube method, evidence was developed suggesting that the blood viscosity would only be substantially decreased when the total lipid intake was less than 40 m daily. McDONALD and EDGILL [56] also observed that the adhesiveness of platelets in patients with vascular disease decreased when a very low fat diet was consumed.

Importance of Circadian Periodicity Changes in Blood Viscosity of Humans

Circadian periodicity changes in blood viscosity and related factors may also influence the cerebral circulation, and contribute to cerebral ischemia [48, 49]. It is known that a majority of vascular accidents of the brain occur during sleep, often considered to be related to a drop in blood pressure. Blood viscosity and factors which influence it such as total proteins, hexosamines, protein bound carbohydrates, sialic acid, and the concentration of red cells in the circulating blood are lowest at this time. After arising, all of these parameters gradually increase to maximum during the middle of the day where they remain until evening when they again decrease. If strenuous exercise for 5 min is undertaken immediately after rising, the mid-day levels of these parameters are reached in a few minutes. No such periodicity was noted for calcium, sodium, or potassium in plasma or red cells.

In all likelihood, shifts of water between tissues and the blood are important causes of these rhythms, but one must also consider the possibility that sequestration and an increase in concentration in the capillary beds of red cells, plasma proteins and other substances closely bound to red cells and plasma proteins, are in part responsible for the circadian changes enumerated above. This would lead to an increase in viscosity of blood and slowing of circulation in the microvasculature of certain tissues, i.e., brain, and thereby contribute to focal ischemia. At the same time, the viscosity of the circulating blood would be reduced. Sequestration of red cells was observed in animals which had received injections of serotonin [23].

Importance of in Vitro Formed Platelet Aggregates

Until now we have considered formation of microemboli *in vivo* by aggregation of blood elements. The introduction into the body of large numbers of such aggregates, formed *in vitro*, is also possible. During storage of blood for transfusion, platelet and leukocyte aggregates form [50, 51, 52]. These aggregates average 25 to 50 micra in diameter, large enough to occlude the conduits of the microcirculation, and massive numbers are often introduced into the body during the use of an extracorporeal circulation for heart and other surgery. Following cardiac surgery focal neurological damage was frequently observed [53]. The symptoms and signs appeared the second post-operative day, became most marked in 5 to 7 days, then receded. Focal electroencephalographic changes, and mental confusion usually accompanied the focal neurological changes. Post-mortem examination of patients who died within 4 days of surgery have revealed platelet microemboli in the microvasculature of various organs of the body [54]. These studies would appear to clearly indicate the possibility that microemboli, formed by aggregated platelets (and leukocytes), are by themselves capable of causing focal cerebral ischemia and hence focal neurological deficits.

Comments

In the discussion so far we have considered a number of factors which can, under certain circumstances, contribute to the development of abnormal adhesiveness of platelets, and thus to their aggregation and to occlusion of the microvasculature of various organs of the body and in particular the brain. It seems unlikely that a young normal circulatory system would be seriously embarrassed by these aggregates with the exception of the lung following severe hypotensive-traumatic "shock". In this circumstance it is now reasonably established from war experiences in Viet Nam that wide spread occlusion of pulmonary vascular conduits can occur in human beings [55].

However, in general the so-called microemboli of "sludge" or adhesive and aggregated platelets probably become effective in an older circulation already compromised by arterial atherosclerosis. In such cases, the presence of intravascular aggregation (red cells or platelets) due to myocardial or pulmonary disease, impaired circulation of an extremity from atherosclerosis, or from traumatic or hypotensive "shock" could precipitate focal cerebral ischemia, or diffuse changes simulating a toxic state. A potentially far more important precipitating series of events could just as well occur among the high fat consuming western populations as the result of either acute or chronic high lipid intake. In all these instances the daily circadian pattern of blood viscosity alteration could play an important role.

Aggregated blood elements would hardly be expected to occlude arteries, but rather arterioles and capillaries. Given a circulation already critically slowed by narrowing of the major arteries, subsequent occlusion or marked slowing of blood flow at the capillary level could produce the degree of ischemia necessary to destroy the function of tissues. If the area involved were large, signs and symptoms of focal ischemia, i.e., hemiparesis, would result. If multiple small ischemic foci resulted one would expect more general symptoms, i.e., mental impairment and confusion to result. It can be anticipated that capillary beds linking several areas, each supplied by a major cerebral or spinal artery ("watershed areas"), would be especially vulnerable to occlusion by aggregated blood cell microemboli [57].

At the moment we are only able to support our hypothesis indirectly, and principally with physiological data mostly from animals. It is well to keep in mind, however, that a rather inconsistent relationship exists in humans between the major cerebral blood vessels which are found occluded, either pre-mortem by arteriography, or postmortem by actual observation, and the area of tissue softening and the functional loss. This inconsistency might be explained in some degree by the hypothesis discussed in this paper.

References

1. BLOCH, E. H.: *In vivo* microscopic observations of the circulating blood in acute myocardial infarction. Amer. J. med. Sci. **229**, 280 (1955).
2. KNISELY, M. H., ELIOT, S. E., BLOCH, E. H.: Sludged blood in traumatic shock. Arch. Surg. **51**, 220 (1945).
3. DAVIS, E., LANDAU, J.: Clinical capillary microscopy. Springfield (Ill.): Charles C. Thomas 1966.
4. SWANK, R. L., DAVIS, E.: Blood cell aggregation and screen filtration pressure. Circulation **23**, 617 (1966).
5. McDONALD, L., EDGILL, M.: Coagulability of the blood in ischemic heart disease. Lancet **1957 II**, 457.
6. HAMPTON, J. R., MITCHELL, J. R. A.: Abnormalities in platelet behaviour in acute illness. Brit. med. J. **1966 I**, 1078.
7. SWANK, R. L., HISSEN, W., BERGENTZ, S. E.: 5-Hydroxytryptamine and aggregation of blood elements after trauma. Surg. Gynec. Obstet. **119**, 779 (1964).

8. — Adhesiveness of platelets and leukocytes during acute exsanguination. Amer. J. Physiol. **202**, 261 (1962).
9. — Hissen, W., Fellman, J. H.: 5-Hydroxytryptamine (serotonin) in acute hypotensive shock. Amer. J. Physiol. **207**, 215 (1964).
10. Hirsch, H., Breuer, M., Kunzel, H. P., Marx, E., Sachweh, D.: Über die Bildung von Thrombocytenaggregaten und die Änderung des Hämatokrits durch komplette Gehirnischämie. Dtsch. Z. Nervenheilk. **186**, 58 (1964).
11. — Gaehtgens, P., Sobbe, A.: Änderungen des Siebungsdrucks nach Ischämie von Gehirn, Extremität und Niere. Pflügers Arch. ges. Physiol. **281**, 191 (1964).
12. Goodman, J. R., Lim, R. C., Blaisdell, F. W., Hall, A. D., Thomas, A. N.: Pulmonary microembolism in experimental shock. Amer. J. Path. **52**, 391 (1968).
13. Hissen, W., Swank, R. L.: Screen filtration pressure and pulmonary hypertension. Amer. J. Physiol. **209**, 715 (1965).
14. Swank, R. L., Edwards, M.: Microvascular occlusion by platelet emboli after shock and transfusion. Microvasc. Res. **1**, 15 (1968).
15. — Hirsch, H., Breuer, M., Hissen, W.: Effect of glass wool filtration on blood during extracorporeal circulation. Surg. Gynec. Obstet. **117**, 547 (1963).
16. French, E. L., Ada, G. L.: Action of the receptor destroying enzyme of V. cholera (RDE) in guinea pigs. Aust. J. exp. Biol. med. Sci. **32**, 165 (1954).
17. Seaman, G. V. F., Swank, R. L.: The influence of electrokinetic charge and deformability of red blood cells on the flow properties of its suspensions. Biorheology **4**, 47 (1967).
18. Hirsch, H., Swank, R. L., Breuer, M., Hissen, W.: Screen filtration pressure of homologous and heterologous blood and electroencephalogram. Amer. J. Physiol. **206**, 811 (1964).
19. Stehbens, W. E.: Early intravascular platelet agglutination. Bibl. anat. (Basel) **9**, 80 (1966).
20. Hellem, A. J.: The adhesiveness of human blood platelets *in vitro*. Scand. J. clin. Lab. Invest. (Suppl.) **12**, 1 (1960).
21. Ditzel, J.: Relationship of blood protein composition to intravascular erythrocyte aggregation. Copenhagen: Christtreus Bogtrykkeri 1959.
22. Bolton, C. H., Hampton, J. R., Mitchell, J. R. A.: Nature of transferable factor which causes abnormal platelet behaviour in vascular disease. Lancet **1967 II**, 1101.
23. Swank, R. L., Fellman, J. H., Hissen, W.: Aggregation of blood cells by 5-hydroxytryptamine (serotonin). Circulat. Res. **13**, 392 (1963).
24. Mitchell, J. R. A., Sharp, A. A.: Platelet clumping *in vitro*. Brit. J. Haemat. **10**, 78 (1964).
25. Swank, R. L., Fellman, J. H.: Blood cell aggregation and biological agents. Bibl. anat. (Basel) **9**, 98 (1967).
26. — The screen filtration pressure in platelet research: Its significance and interpretation. Series Haematologica, Vol. 2, Blood Platelets, 1968.
27. — Hissen, W.: Influence of serotonin on cerebral circulation. Arch. Neurol. **10**, 468 (1964).
28. — Seaman, G. V. F., Hissen, W., Lino, L.: Physico-chemical changes in blood induced by trauma. Surg. Gynec. Obstet. **123**, 251 (1966).
29. Hissen, W., Swank, R. L., Lino, L., Seaman, G. V. F.: Physico-chemical changes in circulating canine blood on exsanguination or administration of histamine. Surg. Gynec. Obstet. **122**, 1003 (1963).
30. Seaman, G. V. F., Vassar, P. S.: Changes in electrokinetic properties of platelets during their aggregation. Arch. Biochem. **117**, 10 (1966).
31. Cullen, C. F., Swank, R. L.: Intravascular aggregation and adhesiveness of the blood elements associated with alimentary lipemia and injections of large molecular substances: Effect on blood-brain barrier. Circulation **9**, 335 (1954).
32. Swank, R. L.: Changes in the blood produced by fat meals and intravenous heparin. Amer. J. Physiol. **165**, 798 (1951).
33. — Effects of fat on blood viscosity in dogs. Circulat. Res. **4**, 579 (1956).
34. — Changes in blood of dogs and rabbits by high fat intake. Amer. J. Physiol. **196**, 473 (1959).
35. Meyer, J. S., Waltz, A. G.: Effects of changes in composition of the plasma on pial blood flow. Neurology (Minneap.) **9**, 728 (1959).

36. Harders, H.: Neue Beobachtungen zum Diätfehler. Verh. dtsch. Ges. inn. Med. **62**, 499 (1956).
37. Williams, A. V., Higginbotham, A. C., Knisely, M. H.: Increased blood cell agglutination following ingestion of fat, a factor contributing to cardiac ischemia, coronary insufficiency and anginal pain. Angiology **8**, 29 (1957).
38. Swank, R. L.: Effect of high fat feeding on viscosity of blood. Science **120**, 427 (1954).
39. — Roth, E. S.: Hemolysis and alimentary lipemia: Effect of incubation, heparin and protamine. Blood **9**, 348 (1954).
40. — Nakamura, H.: Oxygen availability in brain tissues after lipid meals. Amer. J. Physiol. **198**, 217 (1960).
41. — — Convulsions in hamsters after cream meals: Electroencephalograms and available cerebral oxygen. Arch. Neurol. Psychiat. (Chic.) **3**, 594 (1960).
42. Nakamura, H., Swank, R. L.: Electrocardiogram in hamsters after large fat meals. Proc. Soc. exp. Biol. (N. Y.) **105**, 195 (1960).
43. Swank, R. L., Jackson, L.: Electrolyte changes in blood, urine and tissues after lipid and other test meals. Amer. J. Physiol. **204**, 1071 (1963).
44. — Roth, J. G.: Apparatus for measuring relative blood viscosity. Rev. Sci. Instruments **25**, 1020 (1954).
45. — Blood viscosity in cerebrovascular disease: Effect of low fat diet and heparin. Neurology (Minneap.) **9**, 553 (1959).
46. Dintenfass, L.: Consideration of the internal viscosity of red cells and its effect on the viscosity of whole blood. Angiology **13**, 333 (1962).
47. Swank, R. L.: The influence of ecological factors on blood viscosity and sedimentation and on serum cholesterol. Amer. J. clin. Nutr. **19**, 418 (1962).
48. Seaman, G. V. F., Engel, R., Hissen, W., Swank, R. L.: Circadian periodicity of viscosity and of some protein components in circulating blood. Nature (Lond.) **207**, 833 (1965).
49. Halberg, V. F., Engel, R., Swank, R. L., Seaman, G. V. F., Hissen, W.: Cosinor-Auswertung circadianer Rhythmen mit niedriger Amplitude im menschlichen Blut. Phys. Med. Rehab. **7**. Jahrg., Heft 5, Mai 1966.
50. Swank, R. L.: Alteration of blood on storage: Measurement of adhesiveness of "aging" platelets and leukocytes and their removal by filtration. New Engl. J. Med. **265**, 728 (1961).
51. — Porter, G.: Disappearance of microemboli transfused into patients during cardio-pulmonary bypass. Transfusion (Philad.) **3**, 192 (1963).
52. Kunzel, H. P., Hirsch, H.: Über die Entstehung von Aggregaten in ACD-Blutkonserven. Acta haemat. (Basel) **32**, 89 (1964).
53. Singh, N., Carter, C. C., Swank, R. L., Blachley, P.: Relationship between post-cardiotomy delirium, clinical neurology, and EEG abnormalities. J. thorac. cardiovasc. Surg. **54**, 557 (1967).
54. Jenevein, E. P., Weiss, D. L.: Platelet microemboli associated with massive blood transfusion. Amer. J. Path. **45**, 313 (1964).
55. Conference on the pulmonary effects of non thoracic trauma. Washington, D. C.: National Science Foundation 1968.
56. McDonald, L., Edgill, M.: Dietary restriction and coagulability of the blood in ischemic heart disease. Lancet **1958** J, 996.
57. Bartsch, W., Swank, R. L.: Der Effekt von Herzleistung und Blutdruck auf die Hämodynamik der spinalen Durchblutung. Verh. dtsch. Ges. inn. Med. **72**, 1105—1110 (1966).

Clinical Features and Pathogenesis of Cerebrovascular Disease: Hypertension and Hypotension

Hypertension: Permanent or Temporary Cause of Cerebral Infarction and Insufficiency*

J. MARSHALL

The relationship between hypertension and cerebral haemorrhage is too well known to require emphasis. Rather is there a need to depreciate the connection between the two for there is increasing epidemiological evidence that cerebral haemorrhage has been greatly over-diagnosed as a cause of death in the past (HEASMAN and LIPWORTH, 1966; KURLAND, CHOI and SAYRE, 1967). This over-emphasis has been accompanied by a failure to appreciate the important role that hypertension plays in the genesis of cerebral infarction and in the production of transient episodes of cerebral ischaemia. The purpose of this article is to present the evidence bearing upon this important aspect of cerebrovascular disease.

Regrettable though this failure was, its practical consequences were negligible when there was no effective treatment for hypertension. The advent of hypotensive drugs, which can maintain the blood pressure at a relatively constant and desired level, has changed this. It is now incumbent upon us to explore in increasing depth the relationship of hypertension to the genesis of complete and transient strokes.

The Effects of Hypertension on Cerebral Arteries

The effects of prolonged hypertension upon the carotid and vertebral arteries and their intracerebral branches are several; firstly, it aggravates the development of atherosclerosis in the larger vessels and causes it to appear in vessels of smaller calibre than usual. Hypertension is not, of course, the sole factor responsible for the production of atherosclerosis; other influences are clearly at work, discussion of which is beyond the scope of the present article, but the important contribution of high blood pressure as an aggravating factor must not be over-looked.

The second effect of prolonged hypertension is to produce that pathological change which is usually referred to as hypertensive vascular disease. This affects the smaller intracerebral arteries and is characterised by hypertrophy of the medial musculature with subsequent degeneration of the muscle fibres and their replacement by fibrous tissue.

---

* From the National Hospital for Nervous Diseases, Queen Square, London, W. C. 1.

The third effect was described as long ago as 1868 by Charcot and Bouchard, but forgotten until the recent work of Green (1930), Ross Russell (1963) and Cole and Yates (1967). This consists of the development of miliary aneurysms on the short penetrating striate arteries which supply the deep white matter of the cerebral hemispheres. They also appear to a lesser extent upon the long penetrating arteries which supply the subcortical white matter. Age clearly plays a part in the development of these miliary aneurysms, but they are found with much greater frequency in the hypertensive than in the non-hypertensive subject.

The fourth effect of prolonged hypertension is to facilitate the development of the congenital berry aneurysms which are distributed mainly around the circle of Willis and its immediate branches. The predisposing cause of these aneurysms is a congenital defect in the medial coat of the vessel, but hypertension plays a contributory, though not essential, role in their subsequent expansion and rupture, some 60% of patients with ruptured cerebral aneurysms being hypertensive (Crompton, 1966).

The fifth and final effect of prolonged hypertension is, in a minority of cases, to produce the arteriolar necrosis which is associated with the clinical picture of malignant hypertension. It is in the light of these varied effects of hypertension upon the cerebrovascular tree that we can now attempt to interpret some of the events we encounter in patients with hypertension and cerebrovascular disease.

Hypertension and Cerebral Infarction

There has been a widespread tendency in clinical practice to equate the term cerebral infarction with that of cerebral thrombosis and this despite the fact that it is impossible to find the alleged thrombus in many cases of undoubted infarction. There has been an equally widespread tendency to relate the development of the hypothetical thrombus to a state of low blood pressure. This erroneous thinking has led in turn to a failure to appreciate that the majority of cases of cerebral infarction are associated with hypertension rather than with hypotension. Low-Beer and Phear (1961) demonstrated this in the study in which they related their findings at autopsy to the pre-morbid level of blood pressure. Likewise, Prineas and Marshall (1966) in their study of 135 clinically diagnosed cases of cerebral infarction found that the great majority had a blood pressure which was raised to varying degrees above the average for the normal population of the same age and sex (Fig. 1). This elevation of blood pressure was not due to the immediate effects of the stroke, because the blood pressure was that recorded when the patients were convalescent.

The relationship of hypertension to cerebral infarction is, however, by no means simple, as the study of Prineas and Marshall further showed. They found that if they divided the patients according to the degree of elevation of their diastolic blood pressure, those whose pressure, though raised, was below 110 mmHg, presented a picture of a large infarct in the cerebral hemisphere associated with a high incidence of extracranial and intracranial vascular stenoses and occlusions. They contrasted clearly with those patients whose diastolic pressure was 110 mmHg or above for these latter presented the picture of a small lesion situated deep in the cerebral hemisphere and unassociated with stenotic or occlusive disease of vessels.

This study was clinical rather than pathological, but the few cases that came to autopsy confirmed this broad distinction between patients with greatly raised diastolic blood pressure who tended to develop small deep-seated lesions in the white matter and those with a lesser elevation of pressure who developed massive hemisphere infarcts.

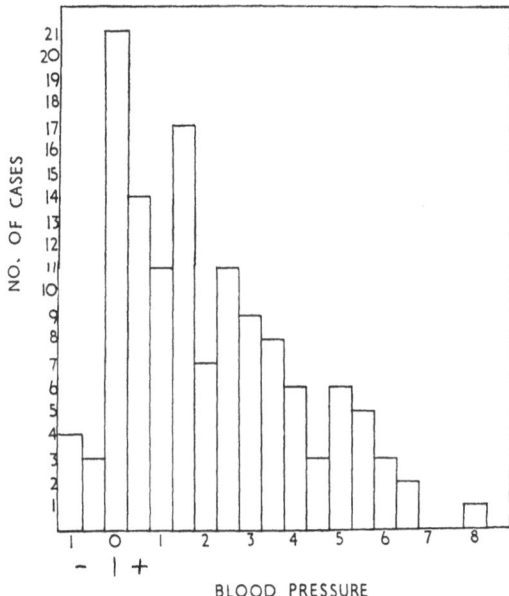

Fig. 1. Histogram of the factor by which the blood pressure of patients with cerebral infarction differed from the mean blood pressure of normal subjects of the same age and sex. (Reprinted from the Brit. Med. J., **1966 I**, 15, by permission of the authors', editor and publishers)

Infarction and Mild Hypertension

This evidence is such as to enable us to frame a tentative hypothesis to be tested in more extensive clinico-pathological studies; it is one which has the merit of forming the basis for a rational approach to the clinical management of patients. In the patients whose diastolic pressure was elevated to less than 110 mmHg, the role of the hypertension had been mainly that of aggravating the development of athero-sclerosis, thus promoting the stenotic and occlusive lesions which were found in both intra- and extracranial arteries. Prevention here can only be by the long-term control of blood pressure before such lesions are formed, along with such other measures as are ultimately found to retard the formation of atherosclerosis.

The management of the infarction in the acute stage in this group must take into account certain pathophysiological facts. When the infarction is due to occlusion or severe stenosis of a vessel, it lies in the centre of the territory of supply of that vessel. The periphery of the territory is preserved because it receives blood via collaterals from neighbouring vascular territories. The factors determining the shift of blood into the ischaemic area are firstly, the pressure differential, the pressure being low in the territory of the occluded vessel and higher in the surrounding vascular territories, and secondly, the vasodilatation which develops in the ischaemic area because of the accumulation of CO_2. It is important that nothing should be done to disturb these relationships; the blood pressure must not be lowered or the pressure differential between the occluded and non-occluded territories may be eliminated; CO_2 should not be administered as it will produce generalized cerebral vasodilatation and so deprive the ischaemic area of its preferential arrangement. General supportive measures are all that can be applied until the convalescent stage is reached.

The management of the infarction once convalescence has been reached must depend upon the state of the patient. Patients whose handicap is so great, dementia so marked, age so advanced or cardiovascular status so poor as to prevent them from gaining appreciable benefit from any intervention in the cerebrovascular situation are best left alone. But for those patients to whom these restrictions do not apply (and there are an increasing number of patients experiencing cerebral infarctions in their prime of life) much can be done. Extracranial vascular stenoses should be relieved by surgery irrespective of whether it is thought they are interfering with blood flow or providing a site for the formation of emboli. Intracranial stenoses beyond the reach of surgery should be treated by anticoagulant therapy. The purpose of this is to prevent a stenosis becoming occluded by thrombus and to prevent it from providing a site for the formation of emboli which may impact in more peripheral parts of the vascular tree.

If stenoses can be removed surgically then subsequently the blood pressure should be reduced so as to retard the development of new lesions. But when it is not possible to remove a stenosis, great care is required. If the elevation of blood pressure is not great (which is the usual state of affairs in this type of case) it is best left alone, though the patient should be under regular observation so that any subsequent more serious rise of blood pressure can be dealt with.

In those exceptional cases in which the stenoses cannot be relieved surgically and the blood pressure is greatly elevated (e.g. diastolic above 110 mmHg) the effects of cautiously lowering the blood pressure should be observed. This is best done as an acute measure with the patient on a tilt-table so that in the event of the fall in pressure producing any untoward effect the patient's head can be immediately lowered and, if necessary, a pressor agent administered. If, however, he tolerates the acute lowering of his blood pressure satisfactorily, then long-term treatment, avoiding drugs which have a marked postural hypotensive effect, should be instituted. The alternative of leaving the patient's blood pressure uncontrolled exposes him to a considerable risk of reduction of life-expectancy through death from cerebral haemorrhage.

Infarction and Severe Hypertension

We can turn now to the second group of patients, namely those who present a picture of a small deep-seated lesion in the white matter, who have little evidence of stenotic or occlusive disease and whose diastolic blood pressure is high, that is above 110 mmHg. These are the patients who, according to our hypothesis, have developed hypertensive vascular disease with medial hypertrophy and the miliary type of aneurysms described by CHARCOT and BOUCHARD. These aneurysms may certainly rupture and give rise to massive cerebral haemorrhage (HERMAN and McGREGOR, 1940) or to a small restricted perivascular haematoma (COLE and YATES, 1967). The sac of the aneurysm may, however, thrombose (COLE and YATES, 1967) and presumably the thrombosis may spread back to involve the parent vessel so giving rise to a small perivascular infarct. Necrosis and removal of the infarcted material gives rise to the small cystic lesions or lacunes which have been the subject of extensive study by MILLER-FISHER (1965). Because these lesions arise in relation to the small penetrating arteries supplying the deep white matter, the clinical deficit may be severe despite their small size. On the other hand, the degree of clinical recovery may be surprisingly great in contrast to the outcome of the massive hemisphere infarcts developing in relation to the occlusive and stenotic lesions already described.

The management of the small deep-seated infarcts remains difficult. Because on clinical grounds it is impossible to distinguish between a small peri-aneurysmal haemorrhage and a small infarct, anticoagulants should not be given. As the striate arteries are end-arteries there is no point in maintaining a high level of blood pressure in order to facilitate the movement of blood from the neighbouring vascular territories to the ischaemic area. Indeed, as the lesion may be a small haemorrhage rather than an infarct, there is much to be said in favour of lowering the blood pressure cautiously in the acute stage in order to reduce the risk of a small haemorrhage expanding into a devastating intracerebral haemorrhage. This is by no means a rare occurrence. When the history of patients who have suffered a massive intracerebral haemorrhage is taken with care, the story of preceding short-lived episodes which may represent the first perivascular leak can often be elicited (MARSHALL, 1964). The physician, however, may not always feel sufficiently confident that he is not dealing with a large hemisphere infarct in which it is important to do nothing to hamper the shift of blood from neighbouring territories to the ischaemic area via collaterals. In this case it is better to wait until the acute phase is passed before intervening. Once convalescence is established, however, the lowering of the blood pressure becomes imperative. Whether reduction of the blood pressure at this stage prevents the formation of further miliary aneurysms is unknown, but it is certainly reasonable to postulate that it will help to prevent the rupture of those already existing.

Hypertension and Congenital Berry Aneurysms

The fact that a high percentage of patients who suffer a rupture of a congenital berry aneurysm are hypertensive has already been mentioned. It is therefore difficult to absolve the raised blood pressure of all responsibility for the formation and rupture of the aneurysm, the devastating effects of which are well known. The effusion of blood into the subarachnoid space is a potent cause of arterial spasm which may be sufficiently severe and sufficiently prolonged to cause infarction in the territory of supply. The surgical manipulations which are necessary for the treatment of aneurysms may also precipitate or aggravate pre-existing spasm and again lead to infarction. Hypertension, therefore, makes a contribution, albeit indirect, to this complex situation.

Once the aneurysm has been successfully treated by surgery and the patient is convalescent, the problem of persisting hypertension may remain. If the aneurysm has been clipped or wrapped, arteries having been left patent, there would seem to be no contraindication to lowering the blood pressure; indeed there is every indication to do so in order to reduce the risks of further rupture. But when the treat-

12*

ment of the aneurysm has involved the clipping of an artery, such as the anterior
cerebral or the carotid artery in the neck, the situation is more difficult. There can be
no doubt that aneurysms can re-expand after clipping of a supplying vessel because
of the development of collateral channels of supply. Though these channels may be
detrimental to the healing of the aneurysm, they may be essential to the part of the
brain originally supplied by the vessel which has been clipped. In this situation much
attention should be paid to the level of the blood pressure. Mild elevations, as with
diastolic pressures of 110 mmHg or less, should be left alone, but the patient watched
carefully. More severe elevations demand treatment, as the risks of hypertension are
probably greater than those of treatment. Here again, observation of the effects of
acute hypotension on a tilt-table is a good method of determining the appropriate
method of treatment.

Hypotension and Cerebral Infarction

Hypertension clearly plays an important role in the genesis of cerebral infarction
in a variety of ways. Much stress has been laid upon this in this article because of
the widespread tendency to ignore or even deny its role in this regard. This stress
must not, however, lead us to neglect the role of hypotension in the production of
cerebral infarction. This is an infrequent cause, [only 7 of the 135 cases in the series
of PRINEAS and MARSHALL (1966) had a blood pressure below that expected for their
age and sex, Fig. 1] nevertheless it is not one to be ignored. The clinical situations
in which cerebral infarction due to hypotension occurs are well recognised. Falls in
cardiac output due to dysrhythmia or myocardial infarction, hypovolaemia from
gastro-intestinal or other haemorrhage, surgical shock, the effect of sedative, tran-
quillising and other drugs upon old people, provide the situations in which the blood
pressure may fall below the level at which the autoregulatory mechanism is effective
in maintaining adequate flow or in which the mechanism itself is damaged. Infarction
may then occur and as the extensive work of ZÜLCH (ZÜLCH and KLEIHUES, 1967)
has shown is situated at the water-shed areas between the vascular territories. This
siting contrasts strikingly with that encountered when a main artery is blocked; in
the latter situation it is the centre of the arterial territory of supply that undergoes
infarction, the water-shed areas being preserved by the collateral supply from the
neighbouring vessels.

The management of this situation is clearly indicated by the mechanism of its
occurrence. The head must be immediately lowered to aid perfusion of the cerebral
vessels and the blood pressure restored by appropriate measures as rapidly as possible.

Transient Hypertensive Crises

The situations described so far have been those in which hypertension has led to
infarction with the persistent neurological deficit of varying severity that this must
cause. Hypertension may also cause ischaemia of sufficiently short duration to produce
the familiar clinical syndrome of the transient ischaemic attack. This it may do in a
variety of ways. One of these is during the developing phase of malignant hypertension. BYROM (1954, 1963) showed that this condition is associated with arteriolar
spasm sufficient to produce ischaemia and cerebral oedema. When the spasm is wide-

spread, hypertensive encephalopathy of the type described by FISHBERG develops with headache, vomiting, fits, papilloedema and depression of consciousness. This phase may be preceded by focal arteriolar spasm giving rise to focal neurological disturbances of the type encountered in transient ischaemic attacks. If there is a spontaneous or therapeutically induced reduction in the level of blood pressure, the transient ischaemic attack may be all that will occur, the fully developed picture of hypertensive encephalopathy never appearing.

Transient episodes of this kind constitute a medical emergency, for the mortality of hypertensive encephalopathy is high; immediate treatment by prompt lowering of the blood pressure is imperative. This measure may be neglected because the patient is already receiving treatment with hypotensive drugs and the transient neurological disturbance is mistakenly attributed to a hypotensive episode rather than to a hypertensive crisis. Proper observation of the patient with frequent measurement of the blood pressure enables this disastrous mistake to be avoided.

Transient Episodes of Other Kinds

Transient disturbances may also develop in relation to changes in the miliary aneurysms of the CHARCOT-BOUCHARD type already described. Leaks from these aneurysms may be of such slight degree as to cause only short-lived disturbance of neurological function. They may precede a more serious rupture leading to a massive cerebral haemorrhage as the history of many patients with this condition shows.

Finally hypotensive episodes may, when sufficiently brief, produce only transient disturbances of neurological function, the blood pressure and blood flow being restored to adequate levels before infarction occurs. Thus hypertension, or its obverse, hypotension, may contribute not only to the production of infarction in one or other of the ways described above, but may also give warning of impending serious pathology by causing short-lived periods of ischaemia insufficient to produce infarction.

The relationship of hypertension to cerebrovascular disease is therefore much more complex than was hitherto envisaged. The old view which equated it largely with the production of massive intracerebral haemorrhage is inadequate. It may be associated with the production of cerebral infarction and of transient ischaemic disturbances in a variety of ways. Greater understanding of its role is of great importance, because the wide range of drugs which are effective in controlling the level of the blood pressure make successful therapy now possible. But this therapy cannot be directed by rule-of-thumb; its proper application demands an understanding of the way in which hypertension contributes to the various manifestations of cerebrovascular disease.

References

BYROM, F. B.: The pathogenesis of hypertensive encephalopathy and its relation to the malignant phase of hypertension. Experimental evidence from the hypertensive rat. Lancet 1954 II, 201—211.
— The nature of malignancy in hypertensive disease. Evidence from the retina of the rat. Lancet 1963 I, 516—520.
CHARCOT, J. M., BOUCHARD, C.: Arch. Physiol. norm. Path. 1, 110, 643, 725 (1868).
COLE, F. M., YATES, P. O.: The occurrence and significance of intracerebral micro-aneurysms. J. Path. Bact. 93, 393—411 (1967).

CROMPTON, M. R.: The pathogenesis of cerebral aneurysms. Brain **89**, 797—814 (1966).
FISHER, C. M.: Lacunes: Small, deep cerebral infarcts. Neurology (Minneap.) **15**, 774—784 (1965).
GREEN, F. H. K.: Miliary aneurysms in the brain. J. Path. Bact. **33**, 71—77 (1930).
HEASMAN, M. A., LIPWORTH, L.: Accuracy of certification of cause of death. General Register Office, Studies on Medical and Population Subjects No. 20. H. M. Stationery Office London 1966.
HERMAN, KATE, McGREGOR, AGNES, R.: Cerebral haemorrhage from rupture of a congenital intracerebral aneurysm in a child. Brit. med. J. **1940 I**, 523—525.
KURLAND, L. T., NUNG WON CHOI, P. H., SAYRE, G. P.: Current status of the epidemiology of cerebrovascular diseases. In: Stroke rehabilitation, basic concepts and research trends. (FIELDS, W. S., SPENCER, W. A., Eds.). St. Louis, Missouri, USA: Warren H. Green 1967.
LOW-BEER, T., PHEAR, D.: Cerebral infarction and hypertension. Lancet **1961 I**, 1303—1305.
MARSHALL, J.: The natural history of transient ischaemic cerebro-vascular attacks. Quart. J. Med. **33**, 309—324 (1964).
PRINEAS, J., MARSHALL, J.: Hypertension and cerebral infarction. Brit. med. J. **1966 I**, 14—17.
ROSS RUSSELL, R. W.: Observations on intracerebral aneurysms. Brain **86**, 425—442 (1963).
ZÜLCH, K. J., KLEIHUES, P.: Neuropathology of cerebral infarction. In: Thule International Symposium on Stroke, April 1966. (ENGEL, A., LARSSON, T., Eds.). Stockholm: Skandia Group, Nordiska Bokhandelns Forlag 1967.

Hypertension and Cerebral Atherosclerosis in Relation to Cerebral Infarction and Cerebral Hemorrhage*

S. Katsuki, T. Omae, N. Okabe, K. Nishimaru, T. Yamaguchi, K. Kimoto, and Y. Nishino

It has been well established that cerebral hemorrhage is almost always associated with high blood pressure unless it is accompanied by cerebrovascular anomaly (saccular aneurysm or angioma) or hematologic disorders, but this is not necessarily so in the case of cerebral infarction. The present communication is concerned with the factors such as hypertension, diabetes mellitus and aging involved in the development of cerebrovascular diseases. For the clinical part of the study, stenotic arterial lesions demonstrable by cerebral angiography were considered in relation to these factors. In autopsy materials, the significance of cerebral atherosclerosis was assessed in the development of the two types of brain lesions.

Material and Methods

One hundred and forty-seven cases with non-embolic, non-hemorrhagic cerebral infarction (cerebral thrombosis) and 34 cases with cerebral hemorrhage other than those caused by vascular malformations and developmental abnormalities, or hematologic disorders, were subjected to the study. All of the cases were those admitted to the Second Department of Internal Medicine, Kyushu University Hospital within one year following the onset of a cerebrovascular episode between the period of March, 1962 and August, 1967. In the case of cerebral infarction there were 113 males and 34 females. The ages ranged from 34 to 85 years (averaging 58) for males and 23 to 79 years (averaging 60) for females. Cases of cerebral hemorrhage consisted of 24 males and 10 females. The ages ranged from 31 to 72 years (averaging 52) for males and from 35 to 69 years (averaging 57) for females. Cerebral infarction was considered when a neurological deficit lasted more than 24 h, the cerebrospinal fluid was clear and contained no red blood cells, and there was no evidence of emboli or intracranial mass lesions. Diagnostic criteria for cerebral hemorrhage followed the description by an *ad hoc* Committee of the advisory council for the National Institute of Neurological Diseases and Blindness, United States Public Health Service [1]. Blood pressures were divided into three groups:

1. Diastolic hypertension; diastolic blood pressure was always or intermittently over 90 mmHg.

---

* From the Second Department of Internal Medicine (Director: Prof. S. Katsuki), Faculty of Medicine, Kyushu University, Fukuoka, Japan.

The study was supported in part by U.S. Public Health Service Grant, NB-03642.

2. Systolic hypertension; systolic blood pressure was always or intermittently over 140 mmHg and diastolic blood pressure always below 90 mmHg.

3. Normal blood pressure; systolic blood pressure remained below 140 mmHg and diastolic blood pressure below 90 mmHg.

Histories of hypertension or receipt of any antihypertensive treatment were checked from the protocols. Diabetes mellitus was considered when postprandial blood glucose levels reached over 170 mg-% and stayed over 140 mg-% for 2 and 3 h after a meal containing 300 gm of rice (100 gm of carbohydrate). In more recent cases, a standard glucose tolerance test was performed. Cerebral angiography was done by inserting a cannula into the subclavian artery in order to visualize the four neck

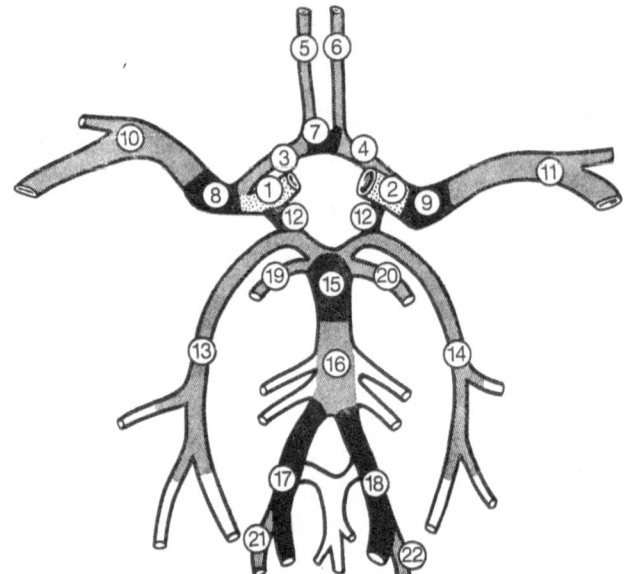

Fig. 1. Twenty-two areas of the circle of WILLIS that are individually coded

arteries, with the few exceptions in which the carotid artery was directly punctured. Fifteen to 20 ml of 60% urografin (Schering, AG) was injected by an Amplatz's automatic injector [2]. Satisfactory visualization of at least three neck arteries was obtained in 84 cases with cerebral infarction and in 13 cases with cerebral hemorrhage, but the analysis was merely made in the former, because in the latter the number of the cases was small.

For pathologic study, 1,016 routine autopsy cases were collected from various pathological institutions in Japan in the years 1962 to 1966 for coding the severity of cerebral atherosclerosis. For coding, the entire circle of WILLIS was divided into 22 clearly defined areas (Fig. 1) and each of these areas graded from 0 to 4 + according to the system described below [3].

0: No involvement.

1 + : Opacity to very thin plaque only in a small part of the vessel without narrowing of the lumen.

2+ : (A) A diffuse thin plaque that does not involve the entire vessel circumference with minimal narrowing of the lumen. (B) A small thick plaque that produces less than 25% narrowing of the lumen.

3+ : (A) A diffuse thin plaque involving the entire circumference with mild narrowing of the lumen. (B) A localized thick plaque producing 25 to 50% narrowing of the lumen.

4+ : (A) A thick plaque involving the entire vessel circumference with moderate to marked narrowing of the lumen. (B) A localized plaque producing over 50% narrowing of the lumen.

The scoring was obtained by adding the totals of all the graded areas and a highest possible score was 88.

Results

1. Blood Pressure in Cerebral Infarction and Cerebral Hemorrhage

In cerebral infarction 86 out of 147 cases (59%) had been known to be hypertensive before admission and 25 (17%) had given no history of hypertension. Either blood pressure had not been measured or no record of blood pressure determinations was

Table 1. *History of hypertension before admission in patients with cerebral infarction and cerebral hemorrhage*

History of hypertension	C. infarction		C. hemorrhage	
	No. of cases	(%)	No. of cases	(%)
Yes	86	(59)	25	(74)
No	25	(17)	5	(15)
Unknown	36	(25)	4	(12)
Total	147	(100)	34	(100)

obtained in the remaining 36 (25%). In cerebral hemorrhage 25 out of 34 cases (74%) had given a history of hypertension, 5 (15%) no history, and the remaining 4 (12%) had no blood pressure records before admission (Table 1). Regarding blood pressure taken during admission, diastolic hypertension was found in 71 (48%) out of 147 cases with cerebral infarction and in 29 (85%) out of 34 cases with cerebral hemorrhage. Most of these cases were accompanied by an increase in systolic blood pressure over 140 mmHg. Systolic hypertension was found in 45 cases (31%) in the former and in 3 cases (9%) in the latter. Thirty-one cases (21%) of the former and 2 cases (6%) of the latter had normal blood pressure (Table 2). It can be assumed from these findings that cerebral hemorrhage is a disease closely linked to hypertensive vascular processes in the brain and this is not necessarily so in the case of cerebral infarction. Average blood pressure at the time of admission was 151/86 mmHg for cerebral infarction and 160/94 mmHg for cerebral hemorrhage. Ten (7%) out of 147 cases with cerebral infarction gave a history of myocardial infarction or angina pectoris but no coronary episode was described in the cases with cerebral hemorrhage.

Among the 84 cases with cerebral infarction in which good visualization of at least three neck arteries was obtained by means of cerebral angiography, 44 cases had either established or intermittent diastolic hypertension, 26 systolic hypertension and 14 had normal blood pressure. Stenotic lesions producing more than 25% narrowing in the diameter, at any portion, of the arteries of the neck or cerebrum were found in 61%, 73%, and 50% of the respective groups of cases mentioned above. Although the cases with systolic hypertension seemed to have more chance to have the lesions,

Table 2. *Blood pressure during admission in patients with cerebral infarction and cerebral hemorrhage*

Type of cases	C. infarction		C. hemorrhage	
	No. of cases	(%)	No. of cases	(%)
Diastolic hypertension (diast. b. p. $\geq$ 90 mmHg)	71	(48)	29	(85)
constant	35	(24)	18	(53)
intermittent	36	(24)	11	(32)
Systolic hypertension (syst. b. p. $\geq$ 140 mmHg and diast. b. p. < 90 mmHg)	45	(31)	3	(9)
constant	19	(13)	1	(3)
intermittent	26	(18)	2	(6)
No hypertension	31	(21)	2	(6)
Total	147	(100)	34	(100)

Table 3. *Vascular lesions disclosed by cerebral angiography related to blood pressure in cerebral infarction (84 cases)*

Type of hypertension	No. of cases	Stenotic lesions present	Location of stenotic lesions[a]			
			Carotid system	Vertebro-basilar system	Extra-cranial	Intra-cranial
Diastolic	44	27 (61)	20 (45)	20 (45)	18 (41)	21 (48)
Systolic	26	19 (73)	16 (62)	11 (42)	13 (50)	16 (62)
No hypertension	14	7 (50)	5 (36)	4 (29)	3 (21)	6 (43)
Total	84	53 (63)	41 (49)	35 (42)	34 (40)	43 (51)

[a] The cases with the lesions present in both territories were counted double.
() Shows percentage of the cases.

the difference between the two groups of systolic hypertension and normal blood pressure was not significant ($0.1 < p < 0.2$). The lesions were in general more frequently found in the intracranial arteries than in the extracranial, and the distribution of the lesions, whether intracranial or extracranial, was not significantly affected by the level of the blood pressure. In the cases with diastolic hypertension and normal blood pressure, the lesions were nearly equally distributed between the carotid and vertebro-

basilar system. In those with systolic hypertension, they were somewhat more fre-
quently found in the carotid system (62%) than in the vertebro-basilar system (42%).
But the difference between the two was not statistically significant $(0.1 < p < 0.2)$
(Table 3).

2. Age as a Factor Related to Angiographic Findings in Cerebral Infarction

Since the cases with systolic hypertension were generally somewhat older than the
others, the age factor was analyzed in relation to angiographic findings. In the systolic
hypertensive group, the average age for cases with stenotic lesions was 65 years and
for those without lesions it was 54 years. In the other two groups, however, the
difference between the cases accompanied by and not accompanied by lesions was
merely 3 to 4 years. The findings indicated that a relatively high incidence of stenotic

Table 4. *Angiographic findings and age related to blood
pressure (84 cases)*

Type of hypertension	Stenotic lesions	No. of cases	Ages averaged
Diastolic	(+)	27	60
	(—)	17	56
Systolic	(+)	19	65
	(—)	7	54
No hypertension	(+)	7	53
	(—)	7	50

lesions in the systolic hypertension group was in part attributable to the fact that
older subjects constituted this group. It was also worth noting that an equal number
of the cases either had or did not have lesions in the normal blood pressure group,
although the average age was 53 years for the former and 50 years for the latter
(Table 4).

When the 84 cases were divided into two groups, below 60 years of age and over,
the lesions were seen in 21 (53%) out of 40 cases for the former and in 32 (73%) out
of 44 cases for the latter. The difference, however, was not statistically significant
$(p < 0.1)$. In regard to the location of the lesions, the difference of incidence between
the two groups was largest for the carotid system $(p < 0.005)$ and not significant for
the vertebro-basilar system $(0.1 < p < 0.2)$. The difference was slightly larger for the
intracranial arteries than for the extracranial arteries (Table 5).

3. Diabetes Mellitus in Cerebral Infarction

Diabetes mellitus was found in 40 (27%) out of 147 cases with cerebral infarction,
in which 19 cases (13%) had been known to be diabetic before the cerebrovascular
episode and the remaining 21 (14%) were detected to have diabetes after admission.
In 34 cases with cerebral hemorrhage 8 (24%) were found to have a diabetic pattern
for the postprandial blood glucose curve, among which only one had been known
to be diabetic before the episode.

Among the 84 cases angiographically studied, stenotic lesions were somewhat more frequently found in diabetics than in non-diabetics, 70% for the former and 61% for the latter. But the difference was not significant ($0.4 < p < 0.5$). There was nothing particular to be mentioned as to the location of the lesions between the two groups (Table 6).

Table 5. *Vascular lesions disclosed by cerebral angiography related to age in cerebral infarction* *(84 cases)*

Age	No. of cases	Stenotic lesions present	Location of stenotic lesions			
			Carotid system	Vertebro-basilar system	Extra-cranial	Intra-cranial
60 >	40	21	14	13	12	15
		(53)	(35)	(33)	(30)	(38)
60 ≦	44	32	27	22	22	28
		(73)	(61)	(50)	(50)	(64)
	P	<0.1	<0.005	0.1 < <0.2	<0.1	<0.02

() Shows percentage of the cases.

Table 6. *Vascular lesions disclosed by cerebral angiography in diabetics and non-diabetics associated* *with cerebral infarction (84 cases)*

	No. of cases	Stenotic lesions present	Location of stenotic lesions			
			Carotid system	Vertebro-basilar system	Extra-cranial	Intra-cranial
Diabetics	23	16	13	10	12	14
		(70)	(57)	(43)	(52)	(61)
Non-diabetics	61	37	28	25	22	29
		(61)	(46)	(41)	(36)	(48)

() Shows percentage of the cases.

4. Pathologic Study

Since reliable recording of ante-mortem blood pressure level was not readily obtained in a considerable number of the autopsy cases, an increase in heart weight was considered to be a manifestation of hypertension. The cases with valvular disease, ischemic myocardial lesions, severe anemia, leukemia and amyloidosis were excluded for these disorders were known to affect heart weight. When the severity score of cerebral atherosclerosis was compared between the carotid system (anterior cerebral, middle cerebral and internal carotid arteries) and the vertebro-basilar system (basilar, posterior cerebral, vertebral, superior cerebellar and posterior inferior cerebellar arteries), it was slightly more advanced in the former than in the latter. In 309 cases of patients over 40 years old and with heart weight less than 350 gm, it was 8.8 ± 0.55 and 6.6 ± 0.44 (mean $\pm$ S. E.) for the carotid and vertebro-basilar system, respectively. In 119 cases with heart weight over 350 gm it was 15.1 ± 1.02 and 11.8 ± 0.83, respectively. In general, the severity score in the cases with heart weight over 350 gm more or less corresponded to that observed in case where the patients were 10 years older and had a heart weight below 350 gm. This was true for both systems. The difference, in relation to heart weight, became smaller with advancing age, particularly in the 8th decade of age and over (Table 7).

In a total of 1,016 cases, the incidence of cerebral infarction and cerebral hemorrhage by age is shown in Table 8. The incidence of cerebral infarction became higher with advancing age, reaching the maximum of 30% for the cases of 8th decade of age

Table 7. *Severity score of cerebral atherosclerosis by age related to heart weight (428 cases)*

Heart weight (gm)		Carotid system					Vertebro-basilar system				
			Decade of age					Decade of age			
		Total	5th	6th	7th	8th ≈	Total	5th	6th	7th	8th ≈
350 ≤	Mean	15.1	8.4	12.4	17.7	21.6	11.8	7.5	9.2	13.7	17.1
	S. E.	1.02	1.98	1.85	1.63	1.95	0.83	1.79	1.42	1.36	1.63
	No. of cases	119	23	33	43	20	119	23	33	43	20
350 >	Mean	8.8	2.7	6.3	10.5	18.8	6.6	2.0	4.7	7.7	14.5
	S. E.	0.55	0.53	0.80	1.04	1.30	0.44	0.49	0.59	0.84	1.10
	No. of cases	309	76	95	82	56	309	76	95	82	56

Table 8. *Incidence of cerebral infarcts and cerebral hemorrhage and no atherosclerotic lesions in the circle of WILLIS by age (1,016 cases)*

Decade of age	No. of cases	Infarct		Hemorrhage		No atherosclerosis in circle of WILLIS	
		No. of cases	(%)	No. of cases	(%)	No. of cases	(%)
1	75	—	—	—	—	75	(100)
2	59	—	—	—	—	54	(92)
3	82	—	—	—	—	73	(89)
4	116	1	(1)	5	(4)	87	(75)
5	164	2	(1)	13	(8)	77	(47)
6	184	10	(5)	21	(11)	44	(24)
7	209	29	(14)	36	(17)	15	(7)
8	86	26	(30)	5	(6)	5	(6)
9	41	11	(30)	1	(2)	1	(2)
Total	1,016	79	(8)	81	(8)	431	(42)

and over, but that of cerebral hemorrhage was highest in the 7th decade of age and decreased thereafter. Although the average incidence of cerebral infarction was not different from that of cerebral hemorrhage, it was evident from these findings that the disease process of the former was closely linked to aging and that of the latter not dependent upon it. The number of the cases with no atherosclerosis in the areas of the circle of WILLIS progressively decreased with age. For instance, in the 4th decade of age, 75% of the cases were free of lesions and the frequency of no lesions was 47%, 24%, 7%, 6% and 2% for the cases of 5th, 6th, 7th, 8th and 9th decade of age, respectively.

Severity of cerebral atherosclerosis related to vascular lesions in the brain 1 cm in diameter and over, is shown in Fig. 2. Cerebral infarcts were found most frequently in the older subjects with a high score cerebral atherosclerosis, but cerebral hemorrhage was seen in the younger subjects with a low score cerebral atherosclerosis as well. For instance, 63 (79%) out of 79 cases with infarcts were encountered in the subjects over 60 years of age with the score of 30 and over but in only 32 (40%) of 81 cases with hemorrhage. A score of less than 10 was not found in cases with infarcts but was found in 10 (12%) out of 81 cases with hemorrhage. Two cases with hemorrhage had scores of zero. When both lesions were present, only cerebral hemorrhage was counted because the hemorrhage was larger than the infarcts and considered more closely related to cause of death. No case was found in which the cause of death was con-

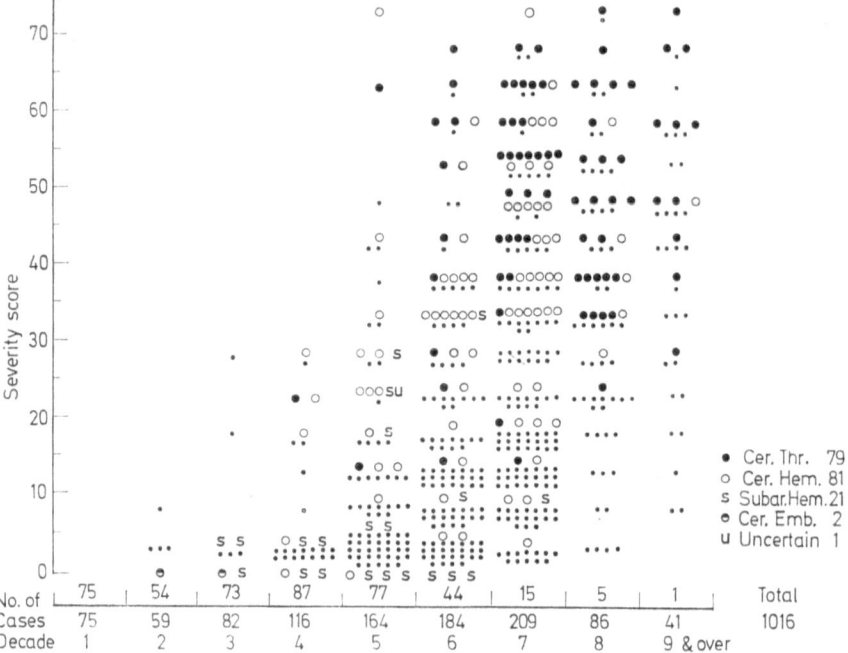

Fig. 2. Severity of cerebral atherosclerosis in relation to various types of vascular lesions in the brain

sidered to be the infarction rather than the hemorrhage when both lesions were present.

Discussion

During the $5^1/_2$ years beginning March, 1962, the total admission of cerebral infarctions to our ward was over four times more common than that of cerebral hemorrhages. Although this ratio of admissions may not have represented an average incidence of cerebral infarction versus cerebral hemorrhage, an increasing trend in the frequency of cerebral infarction in Japan has recently been pointed out [4]. The present report was attempted in order to determine the significance of hypertension and cerebral atherosclerosis in the development of cerebrovascular diseases.

Diastolic hypertension was definitely more frequently associated with cerebral hemorrhage than with cerebral infarction in our clinical series. Blood pressure was normal or systolic blood pressure was merely elevated in about a half of the cases with cerebral infarction. This was well in accordance with the results obtained by our prospective population survey in HISAYAMA [5]. Angiographically demonstrable stenotic lesions producing more than 25% narrowing in diameter were found in 63% of the cases with cerebral infarction. The incidence of the lesions was highest in the cases of systolic hypertension and was presumably due to the relatively large number of older subjects constituting this group. Although stenotic arterial lesions were somewhat more frequent in cases with cerebral infarction accompanied by diabetes mellitus than in those of non-diabetics, the number of the cases was still too small to draw a definite conclusion. The frequency of known diabetes mellitus before cerebrovascular episode was definitely more common in cerebral infarction than in cerebral hemorrhage. As to the distribution of the angiographically demonstrable arterial lesions, the carotid and the vertebro-basilar system were equally affected in the diastolic hypertension group, but carotid system seemed to be somewhat more frequently involved than vertebro-basilar system in the systolic hypertension group. The incidence of arterial lesions was significantly higher in cases where the patients were 60 years of age than in those below 60. The incidence differences were largest in relation to the carotid system and not significant in relation to the vertebro-basilar system.

Frequency and severity of the atherosclerotic process definitely increased with age and also in the presence of hypertension. When a heart weight of more than 350 gm was selected for use as the criterion for the probable presence of hypertension, the severity of cerebral atherosclerosis in the hypertensive subjects (with the exception of those of the 7th decade of age) was more advanced than in normotensive subjects who were 10 years older. This held true both in the carotid and the vertebro-basilar systems. Although the atherosclerotic changes were generally more advanced in the carotid than in the vertebro-basilar system, the difference between the hypertensive and non-hypertensive subjects was considered nearly the same in both systems. Therefore, it could not be said which system was more easily affected in the presence of hypertension. The frequency of no atherosclerosis in the areas of the circle of WILLIS steadily decreased with age. It was, however, worth noting that atherosclerotic changes began to appear as early as in the second decade of age and also there were a few cases with no atherosclerosis even in the 8th decade of age and over. These findings were quite consistent with those reported by BAKER and IANNONE [6, 7]. The incidence of cerebral infarction progressively increased with age reaching the maximum of 30% at the 8th decade of age and over, but cerebral hemorrhage more frequently occurred below this age. It was also shown that cerebral infarcts were most frequently associated with severe cerebral atherosclerosis and cerebral hemorrhage occurred rather independently of these changes.

Summary and Conclusions

1. Diastolic hypertension was observed in most of the cases with cerebral hemorrhage and in about a half of the cases with cerebral infarction.

2. Systolic blood pressure alone was elevated in a considerable number of cases with cerebral infarction in which angiographically demonstrable stenotic arterial

lesions were more frequently found. Presumably this was because of the relatively large number of older subjects constituting this group.

3. The overall incidence of stenotic arterial lesions was 63% for the cases with cerebral infarction. Frequency of their distribution was considered in relation to the type of hypertension, diabetes mellitus and aging.

4. Hypertension, as indicated by heart weight of more than 350 gm, apparently accerelated the atherosclerotic process in the carotid and vertebro-basilar systems to nearly the same degree.

5. The frequency of cerebral infarction increased with age in connection with advancing cerebral atherosclerosis, but cerebral hemorrhage occurred rather independently of the severity of cerebral atherosclerosis.

References

1. A report: A classification and outline of cerebrovascular diseases. Neurology (Minneap.) **8**, 1 (1958).
2. AMPLATZ, K., HARNER, R.: A new subclavian artery catheterization technic. Preliminary report. Radiology **78**, 963 (1962).
3. RESCH, J. A., BAKER, A. B.: Etiological mechanisms in cerebral atherosclerosis. A preliminary study of 3,839 cases. Arch. Neurol. (Chic.) **10**, 617 (1964).
4. KATSUKI, S., HIROTA, Y.: Current concept of the frequency of cerebral hemorrhage and cerebral infarction in Japan. In: Cerebral vascular diseases, p. 99. (MILLIKAN, C. H., SIEKERT, R. G., WISNANT, J. P., Eds.). New York and London: Grune and Stratton 1966.
5. — Cerebrovascular diseases in the Japanese. Jap. J. Clin. Exp. Med. **44**, 2129 (1967) (in Japanese).
6. BAKER, A. B., IANNONE, A.: Cerebrovascular disease. I. The large arteries of the circle of Willis. Neurology (Minneap.) **9**, 321 (1961).
7. — — Cerebrovascular disease. II. A study of etiologic mechanisms. Neurology (Minneap.) **11**, Pt. II: 23 (1961).

Arterial Thrombosis and Microembolism

R. W. Ross Russell*

While the role of platelets in haemostasis has been recognised for over a century [1], the importance of platelets in arterial thrombosis has received insufficient attention, undue emphasis being placed on the coagulation mechanism. It is helpful to regard the formation of the white head of the arterial thrombus as an extension of the processes involved in the haemostatic plug. When a small artery is injured the initial stimulus to thrombus formation appears to be the exposure of subendothelial collagen [2]. Circulating platelets when exposed to collagen are absorbed on to the surface, swell and release many of their constituents into the surrounding blood. Among these are serotonin, which may cause local vascular constriction, and adenosine diphosphate (A.D.P.) [3]. A.D.P. causes a progressive accumulation of platelets at the site of injury which may eventually fill the entire vessel. In contrast to the permanent disruptive change produced by exposure to collagen (platelet adhesion), the reactions produced by A.D.P. are reversible and result in the formation of a loose aggregate, the platelets in the interior of the mass retaining their morphological identity, including electron density, and most of their organelles. Fibrinogen is also necessary for the platelet aggregation produced by A.D.P. but there is no evidence that fibrin binds the platelets together, aggregation being essentially due to the action of A.D.P. Even in a freshly formed platelet thrombus, however, strands of fibrin are found interspersed with platelets and red cells on the surface.

The second stage of haemostasis involves the transformation of a temporary platelet aggregate into a permanent plug. This change is brought about by simultaneous rapid activation of the blood coagulation mechanism, finally resulting in the transformation of fibrinogen to fibrin under the action of thrombin. There are two pathways by which prothrombin is converted to thrombin. The intrinsic pathway is initiated by the conversion of Hageman factor (Factor XII) into its active form by contact with collagen fibres. A cascade of reactions then ensues culminating in the formation of factor X which, in the presence of coagulation factor V and platelet lipids, converts prothrombin to thrombin [27]. A second mechanism, known as the extrinsic pathway, depends upon the release of tissue thromboplastin from damaged vascular walls. This directly activates factor X which converts prothrombin to thrombin as before. Thus, in a vascular injury both extrinsic and intrinsic pathways contribute to the evolution of thrombin. In addition to its action on fibrinogen, thrombin exerts a direct action on platelets resulting in aggregation, swelling and release of granules, and the transformation of a loose platelet mass into a densely adherent plug.

There are many similarities between this process and the formation of a white arterial thrombus. The initiating factor is again the reaction of circulating platelets with an abnormality in the vascular wall, in most cases an atheromatous plaque or a

* St. Thomas' Hospital, Department of Neurology, London.

bare area on the surface of a diseased artery [4, 5]. Platelet-collagen interaction or possibly adhesion of platelets to basement membrane is followed by a progressive growth of a fibrin-platelet thrombus. As the mural thrombus continues to grow, circulation in the area is retarded and this aids the local activation of blood coagulation systems resulting in areas of fibrin and red cell deposition [6]. These areas then become coated with a fresh layer of platelets forming a mixed or laminated thrombus. In a fresh thrombus of this kind the platelets in the depths of the mass may again be intact, while those on the surface are degranulated and bound together by fibrin. Most arterial thrombosis takes place at areas of pre-existing damage, particularly ulcerating atheroma, but blood flow factors are also important in influencing the site of thrombosis. Vascular injury results in more extensive thrombosis if the injury takes place at a junction of the vessel [7] and it is possible that turbulence causes damage to erythrocytes and platelets with local release of A.D.P. favouring platelet aggregation at these sites.

At the periphery of a white thrombus the various forces, both chemical and structural, tending to aggregate platelets are counteracted by the disruptive forces of the blood stream. The effect of A.D.P. on platelet aggregation is lost when the A.D.P. is dephosphorylated to A.M.P. and since the source of A.D.P. is limited, aggregates tend to dissolve unless a stabilising factor such as fibrin is present. This is suggested by experiments on animals with spontaneous or induced coagulative defects where it has been shown that platelet thrombi are unstable and haemostasis is defective [8]. If a mural thrombus is observed in vivo, small platelet masses may be seen being dislodged continuously from the surface only to be replaced by fresh platelets arriving in the blood stream. Larger masses of fibrin and platelets are frequently dislodged as emboli and sometimes the entire vascular plug may be swept forward to impact in distal vessels [9]. Once the fibrin platelet mass has been dislodged from the original site of collagen-platelet interaction, it ceases to grow and tends to fragment, appearing in many cases to pass completely through the capillary circulation. The breakup of thrombus is accelerated by a number of mechanisms such as specific inhibitors of A.D.P., activation of blood fibrinolytic enzymes, local consumption of blood coagulation factors and possibly by hepatic clearance factors [6]. If the fibrin-platelet mass is not swept off as an embolus, it becomes adherent to the surface of the atheromatous lesion, may become organised and finally amalgamated into the plaque [10].

Although embolisation of fibrin-platelet thrombus has been noticed by many observers studying mural thrombosis, the possibility that emboli of this kind might be an important cause of ischaemia appears to be relatively new. From the behaviour of the emboli and the fact that they tend to fragment and pass through smaller vessels, it might be expected that ischaemia would be temporary and perhaps insufficient to lead to infarction of tissue. There are a number of reasons for supposing that such a process would be most noticeable when it affects the cerebral circulation. The brain requires a large constant blood supply and although it is protected by anastomotic arrangements against large vessel occlusion there are relatively few anastomoses between small intracerebral arteries. The central retinal artery is particularly vulnerable in that there are no connexions between the various branches of the central artery above a capillary level. Even a temporary disturbance of flow in such a situation might lead to a state of reversible neuronal damage which rapidly recovered only to recur if a subsequent embolus took the same pathway.

It is possible that the symptoms of cerebral ischaemia which arise in association with carotid artery disease may be due to the dislodgement of microemboli of this kind as well as to the effects of arterial obstruction. The clinical evidence for embolism is strongest in cases of recurrent temporary ischaemia and may be summarised as follows:

1. Transient ischaemic attacks are associated with carotid stenosis rather than with carotid occlusion. Furthermore, transient ischaemia may be cured spontaneously when a carotid stenosis becomes a complete occlusion, thus bringing to an end the possibility of embolism [10].

2. Though transient ischaemic attacks are often stereotyped, repeatedly affecting the same territory, it is possible that haemodynamic factors may direct numbers of similarly sized emboli along the same route. Ischaemia also occurs in widely separate vascular territories, for example the retina and the ipsilateral hemisphere on different occasions.

3. If endarterectomy specimens are examined from cases of recent transient ischaemia, it is found either that the surface of the atheromatous lesion is covered with fresh thrombus, or else that an atheromatous ulcer is present with the exposure of cholesterol containing material to the blood stream. Smooth areas of narrowing or fibrotic nonulcerating lesions are not found in cases of recent transient ischaemia [10].

4. Angiographic studies show evidence of blockage of small intracerebral arteries in the territory served by a stenosed carotid vessel [11].

5. The procedures of endarterectomy [12], administration of anticoagulants [13] or ligation of vessels have all been shown to be effective in abolishing attacks of transient ischaemia.

6. Observations on the retinal circulation have demonstrated the presence of emboli in many cases of carotid artery disease. On a few occasions fragments of white thrombus have been described traversing the retinal artery circulation during attacks of monocular blindness [14, 15]. The visual field loss has been shown to correspond to the area of retina rendered ischaemic. The emboli tended to fragment and appeared to pass through the capillaries resulting in restoration of normal vision. In at least one of these cases, the source of the embolus appeared to be a mixed thrombus originating in the carotid artery and extending into the ophthalmic artery [15]. In other fatal cases, the emboli in retinal arteries were shown to be composed principally of platelets [16] or of lipid material [17]. The periods of occlusion of the artery were often relatively brief, lasting only a few minutes, and it seems unlikely that small fibrin platelet emboli would cause permanent ischaemic damage except possibly in tissues such as the retina with a high metabolic rate and an end-arterial blood supply.

Fragmentation is not confined to microemboli; larger masses of mixed thrombus originating in the left atrium are less likely to enter the retinal circulation but have been studied angiographically after impaction in the carotid and middle cerebral artery and they have been shown to fragment and disappear over the course of a few days, although in this case the period of ischaemia was long enough to cause permanent cerebral damage [18]. The commonest microemboli to lodge in the retinal circulation are undoubtedly small fragments of atheromatous debris with a high content of cholesterol [19]. These have a characteristic refractile appearance in retinal vessels and tend to impact there for longer periods causing an inflammatory reaction

13*

in the vessel wall. Obstruction is seldom complete due to the flat shape of cholesterol crystals [20]. Emboli of a similar nature from aorta and carotid arteries also occur in the cerebral circulation and if widespread may give rise to multiple blockade of small intracerebral blood vessels and cerebral infarction [21].

It must be stressed that not all minor or shortlived episodes of cerebral ischaemia are due to microemboli. Many are due to infarcts produced by local occlusion of intracerebral arteries affected by hypertension, diabetes or arteriosclerosis. Recovery from a small localised infarct may be surprisingly rapid. Other transient attacks, especially when repetitive, are due to failure of collateral blood supply consequent on arterial hypotension, cardiac dysrhythmia or diversion of blood to other areas. Restoration of blood flow leads to a resumption of neuronal activity without structural damage (cerebrovascular insufficiency of Denny Brown) [22]. If, however, it is accepted that a proportion of cerebral ischaemic episodes are due to microembolism, with or without fragmentation, a number of clinical implications follow.

1. Atheromatous lesions in the carotid artery are significant both as a potential obstruction to blood flow and as a source of microemboli. In the latter context the degree of stenosis is unimportant. In some patients marked irregularity of the artery walls exists over most of the length of the vessels, to an extent which makes reconstructive surgery impractible. In many other patients atheromatous change is localised to the area of the bifurcation where the vascular wall appears pitted and irregular, the remainder of the vessel being smooth and regular. Localised areas of ulcerating atheroma may be removed surgically.

2. Medical treatment should aim at the prevention of mural thrombosis on the surface of atheromatous lesions particularly at the initial stage of collagen-platelet interaction. A number of drugs mostly with anti-inflammatory properties such as phenylbutazone and aspirin interfere with this reaction, possibly by preventing release of A.D.P. from platelets [23]. The coagulation mechanism is unaffected.

Of the drugs currently in use, heparin in high dosage has been shown to be effective in preventing platelet thrombosis on foreign surfaces [24] while Coumadin drugs have no such action. Heparin appears to interfere with platelet surface adhesion in addition to its anticoagulant properties. Although in vivo white thrombus continues to form at sites of vascular injury after heparin treatment, the subsequent emboli are smaller and more friable, probably because they lack the stabilising influence of fibrin [9]. Intravascular platelet aggregation may be inhibited altogether by enzyme poisons or by drugs blocking the formation or release of A.D.P. [25] and recent experiments with low molecular weight dextrans have shown that platelet adhesiveness to vascular protheses in vivo is reduced. These observations suggest that the ideal anti-thrombotic agent combining an inhibitory action on platelet adhesiveness and platelet aggregation with anticoagulant properties may shortly be developed. At the moment heparin in high dosage appears to come nearest to the ideal and should be administered to patients with clinical evidence of transient ischaemia due to fragmentary microemboli.

No beneficial action can be expected from anti-thrombotic agents in patients with atheromatous embolism where the emboli are composed of fragments of diseased intima but in this event the embolus acts as an intravascular foreign body and as a locus for further thrombosis and embolism. Secondary thrombosis of this kind may be amenable to medical treatment.

References

1. WHARTON-JONES, W. T.: On the state of the blood and blood vessels in inflammation. Guy's Hosp. Rep. **2**, 71 (1851).
2. HOVIG, T.: Release of a platelet-aggregating substance (adenosine diphosphate) from rabbit blood platelets induced by saline extract of tendons. Thrombos. Diathes. haemorrh. (Stuttg.) **9**, 264 (1963).
3. BORN, G. V. R.: Aggregation of blood platelets by adenosine diphosphate and its reversal. Nature (Lond.) **194**, 927 (1962).
4. CONSTANTINIDES, P.: Plaque fissures in human coronary thrombosis. J. Atheroscler. Res. **6**, 1 (1966).
5. FRENCH, J. E.: Atherosclerosis in relation to the structure and function of the arterial intima with special reference to the endothelium. Int. Rev. exp. Path. **5**, 253 (1966).
6. DEYKIN, D.: Thrombogenesis. New Engl. J. Med. **276**, 622 (1967).
7. GEISSINGER, H. D., MUSTARD, J. F., ROWSELL, H. C.: The occurrence of microthrombi on the aortic endothelium of swine. Canad. med. Ass. J. **87**, 405 (1962).
8. HOVIG, T., ROWSELL, H. C., DODDS, W. J., JORGENSEN, L., MUSTARD, J. F.: Experimental haemostasis in normal dogs and dogs with congenital disorders of blood coagulation. Blood **30**, 636 (1967).
9. HONOUR, A. J., RUSSELL, R. W., ROSS: Experimental platelet embolism. Brit. J. exp. Path. **43**, 350 (1962).
10. GUNNING, A. J., PICKERING, G. W., ROBB-SMITH, A. H. T., RUSSELL, R. W. ROSS: Thrombosis of the internal carotid atery and subsequent embolism. Quart. J. Med. **33**, 155 (1964).
11. CRONQVIST, S.: Total angiography in evaluation of cerebrovascular disease. Brit. J. Radiol. **39**, 805 (1966).
12. DEBAKEY, M. E., CRAWFORD, E. S., COOLEY, D. A., MORRIS, G. C., GARRETT, H. E., FIELDS, W. F.: Cerebral arterial insufficiency: one to eleven years results following arterial reconstructive operation. Ann. Surg. **161**, 921 (1965).
13. SIEKERT, R. G., MILLIKAN, C. H., WHISNANT, J. P.: Anticoagulant therapy in intermittent cerebrovascular insufficiency. J. Amer. med. Ass. **176**, 19 (1961).
14. FISHER, C. M.: Observations of the fundus oculi in monocular blindness. Neurology (Minneap.) **9**, 333 (1959).
15. RUSSELL, R. W. ROSS: Observations on the retinal blood vessels in monocular blindness. Lancet **1961 II**, 1422.
16. MCBRIEN, D. J., BRADLEY, R. D., ASHTON, N.: The nature of retinal emboli in stenosis of the internal carotid artery. Lancet **1963 I**, 697.
17. COGAN, D. G., KUWABARA, T., MOSER, H.: Fat emboli in the retina following angiography. Arch. Ophthal. **71**, 308 (1964).
18. DALAL, P. M., SHAH, P. M., AIYER, R. R.: Arteriographic studies of cerebral embolism. Lancet **1965 II**, 358.
19. HOLLENHORST, R. W.: Ocular manifestations of insufficiency or thrombosis of the internal carotid artery. Trans. Amer. ophthal. Soc. **56**, 474 (1958).
20. RUSSELL, R. W. ROSS, SANDERS, M. D.: A study of retinal vascular occlusions using fluorescein angiography. Lancet **1966 II**, 821.
21. MCDONALD, W. I.: Recurrent cholesterol embolism as a cause for fluctuating cerebral symptoms. J. Neurol. Neurosurg. Psychiat. **30**, 489 (1967).
22. DENNY-BROWN, D.: Recurrent cerebrovascular episodes. Arch. Neurol. Psychiat. (Chic.) **2**, 194 (1960).
23. MUSTARD, J. F., GLYNN, M. F., NISHIZAWA, E. E., PACKHAM, M. A.: Platelet surface interactions: relationship to thrombosis and haemostasis. Fed. Proc. **26**, 106 (1967).
24. GOTT, V. L., WHIFFEN, J. D., DUTTON, R. C.: Heparin bonding on colloidal graphite surfaces. Science **142**, 1297 (1963).
25. HONOUR, A. J., MITCHELL, J. R. A.: Platelet clumping in vivo. Nature (Lond.) **197**, 1019 (1963).
26. BONCHEK, L. I., BRAUNSWALD, N. S.: Modification of thrombus formation on prosthetic valves by the administration of low molecular weight dextran. Ann. Surg. **165**, 200 (1967).
27. MACFARLANE, R. G.: Enzyme cascade in blood clotting mechanism and its function as a biochemical amplifier. Nature (Lond.) **202**, 498 (1964).

Local and Relative Hypotension as the Cause of Cerebrovascular Accidents

A. Pierach*

The concept of an "angiospastic" pathogenesis of c.v.i. has lost its basis as a result of experimental and clinical evidence. On the contrary, the decrease of organ blood volume and deficiency of the general circulation—however induced—may be of greater importance when discussing brain or coronary infarction, Sheehan's syndrome or the disturbances of circulation of other organs.

It seems to be the hemodynamic principle of a *local* and *relative* hypotension of the cerebral arteries which leads to a focal ischemia. This is now well known under the name of cerebrovascular insufficiency.

Our term of "local hypotension" is a clinical term, describing the vascular changes in a particular organ. Experimentally, it is a well proved fact that an area beyond an arteriosclerotic occlusion or stenosis which is without sufficient collateral circulation may have a local, hemodynamically insufficient circulation while the surrounding brain may still be sufficiently supplied. (See the recent investigations on "the cerebral steal".)

The general circulatory factor introducing the local hypotension may have different causes such as an orthostatic disturbance, circulatory collapse, bleeding, circadian or a drug induced hypotension.

This hypotension may be "relative" and be seen clinically in a labile hypertensive where a drop to a systolic blood pressure of 150 mmHg may already have given rise to local circulatory insufficiency in a territory distal to an arteriosclerotic stenosis. Such a pathogenesis is somewhat similar to the "relative" hypoglycemia where a sudden drop to 150 mg-% in a diabetic may lead to hypoglycemic symptoms (MEY-THALER).

The brain with a stenosed sclerotic vascular system seems to need its "specific" pressure in order to maintain a sufficient circulation, which under normal physiological conditions is controlled by so many different mechanisms. Even if normally—according to OPITZ and M. SCHNEIDER and to M. SCHNEIDER and his school— the lowering of the systemic pressure down to 70 mmHg is possible before symptoms of hypoxemia arise, this is different from the case of an increased c.v.r., where a sudden drop to 120 or 100 mmHg (BERNSMEIER, FINNERTY et al.) may give rise to an ischemic lesion, which may become irreversible and induce a brain or coronary infarct. This may be particularly so if another factor comes into the play at the same time i.e. anemia, hypoglycemia or other changes in the blood content. These factors may be super-

* Prof. Dr. med. A. Pierach, Konitzkystift, 6350 Bad Nauheim.
Abbrevations: c.v.i. — cerebrovascular insufficiency; c.v.r. — cerebrovascular resistance; c.v.a. — cerebrovascular accidents.

imposed and reinforce the primary hemodynamic lesion and thereby make it irreversible. (One has to be cautious—by the way—in the interpretation of blood pressure values taken on obese patients, since these are usually "too high" by the modalities of our usual blood pressure cuff!)

We have to take into account that there is no such fixed value as a "basic" blood pressure, but that this consists of relative data, depending on many factors. In 1921 C. MÜLLER found that the blood pressure had its lowest value between 2 and 4 o'clock in the morning and this has been recently reinvestigated by ZÜLCH and V. HOSSMANN (1967) who wanted to explain these c.v.a. at rest ("midnight stroke") as being partly a result of this "low value" in the general blood pressure.

It should be noted that this nocturnal drop in the blood pressure in the early morning is much more marked in the "hypertensive", a fact mentioned long ago by MENZEL and recently stated by RICHARDSON, FENTON, STOTT and PICKERING. Thus the night rest may be dangerous to some hypertensives with particular stenosing changes in the cerebral arteries. Exact values, however, should be taken only by intra-arterial measuring, as we have recommended for some time. This fact has been realized in the intensive care units of many hospitals.

We are indebted to the team of PICKERING cited above for the circadian curves of normals and diseased. These have convinced us once more of the labile values of a 24 h blood pressure curve. Such nocturnal drops of the blood pressure to 100 to 150 mmHg have been seen in patients whose usual values are around 240/160 mmHg. Yet, the Riva-Rocci method does not seem to suffice to "catch" sudden drops as for instance during rising or on the tilting table. These falls, however, can be "read" sufficiently by intra-carotid readings (LOMAN et al., 1936) and confirms our experience as those of MECHELKE and of BACHMANN.

"Local" hypotension may be only sufficiently and primarily measured in organs or body parts whose stenosed circulation induces corresponding "low" values and often here the blood pressure reading may be too late because symptoms are already present and the "low" has been passed already as demonstrated above. This "local" reading is not yet possible as a quick "clinical" method. It does not seem necessary to cite here all the pertinent experiments which are in good correlation with our concept of "local hypotension". We would like to mention only CORDAY's experiments in stenosing the carotid artery. Our concept of the pathogenesis of ischemia of the brain has changed much, particularly since pathology and radiology have shown that infarcts may arise with "open" arteries (though stenosed) and that this is also true for the coronary circulation.

Clinical Observations Associated with "Local" and "Relative" Hypotension

It may be found upon sudden tilting of juvenile hypertensives that the blood pressure drops to 20 to 30 mmHg max. just before collapse. Yet these very low values are present only for a fraction of a minute and usually escape pressure recordings taken by the blood pressure cuff. This test on the "tilting table" may be far more delicate if done with aged hypertensives. Thus, LOMAN et al. have been able to register a blood pressure drop from 200/120 to 70 mmHg max. (intracarotid reading). Such a drop may be also induced for a longer time by nitroglycerines.

A case of orthostatic cerebrovascular ischemia has been seen by my colleague, Prof. Pia. It concerned a man of 50 years of age who suffered practically no other lesion following a road accident. Upon his first attempt to rise, an orthostatic collapse with Jacksonian fits of the left arm and face took place. This stopped immediately when he was put into a horizontal position. Intermittent unilateral changes of reflexes and intermittent EEG changes were observed. The clinical diagnosis was local cerebral contusion provoked by a locally deficient circulation during the collapse.

It may seem superfluous to recall here the experience of our generation at the time when blood-letting was one of the main procedures in internal medicine. The severe lesions described in the literature when blood-letting of the aged was customary are abundant. Yet, even in a boy of 2 years, an agnostic type of amaurosis was observed which lasted for 4 weeks after a traumatic collapse with great loss of blood (Wold).

To take up the comparison again, we saw practically the same type of amaurosis in a 4 year old diabetic, in whom we had great difficulty in regulating the blood sugar. This type of visual blindness lasted 8 h after a drop of the increased blood sugar to (though still above normal) 140 mg-% (see above "relative hypoglycemia").

As another example of c.v.i. or infarct induced by sudden drop of blood pressure, the history of a 46 years old patient may be cited.

He suffered from an aortic coarctation. There was increasing respiratory insufficiency and circulatory disturbances and, because of these and at his personal request, an operation was performed, although he was warned of the possible consequences to his circulation.

Local blood pressure readings: right upper extr. 225/100, left 210/100, max. systolic values in the legs 90 mmHg. Postoperatively, the systemic pressure (intra-arterial reading) was around 132/87. This must have been a "relative hypotension" to the patient. Shortly after the operation, motor and sensory deficiences were noticed on the left side followed by loss of consciousness and death of the patient on the 2nd day.

Autopsy findings: Fresh brain infarct in the stage of "softening" in the right 1st and 2nd frontal convolutions with hemorrhagic softening in the gyrus cinguli, caudate nucleus and putamen. There was severe arteriosclerosis of the brain vessels which was most marked in the sylvian territory.

Anti-Hypertensive Treatment

It is well known that too drastic a lowering of the blood pressure in "fixed" hypertension can induce severe brain lesions (Prineas and Marshall; Spatz; Bernsmeier). It may be particularly dangerous if circadian "lows" are superimposed on the induced lowering of the blood pressure. This has been stressed by Zülch and V. Hossmann.

This pathogenetic "co-operation" of two or several factors (Zülch, 1966) may also be valid for the coronary circulation alone or for both the coronary and cerebral circulations ("cardio-cerebral syndrome" of the Italian authors). Usually the general deficiency of circulation is the *primary* symptom; however, a differential diagnosis may be difficult as demonstrated in the following patient.

Case history: An engineer, aged 80 years, had been working hard in his profession of installing engines and had made a car trip to the place where he worked. He had visitors till late in the night. Finally, he collapsed. There was no history of heart trouble and his blood pressure was said to have been normal. The patient was pale, sweating, had cold extremities and his blood pressure was not measurable. Hemiparesis followed and the patient died a few hours later. Here the sequence of the events was probably: collapse—coronary infarct with deterioration of the circulation—brain infarct.

Induced Hypotension during Operation

Here I would like only to point to BODECHTEL's description of a patient with induced hypotension to 55 mmHg (brain circulation: 22 cc/min/cc according to KETY-SCHMIDT) and add the results of the inquiry of HAMPTON and LITTLE which stated that in 553 cases after operation, 20 were diagnosed as a "cerebral thrombosis" (infarct ?) and two had a retinal thrombosis. Out of 46 patients who died, 11 had a cerebral thrombosis or anoxia.

Other factors which additionally influence the cardio-vascular systems are: paroxysmal tachycardia, hypersensitive carotid sinus reflex, Stokes-Adams syndrome and the dumping syndrome. Most of these can be illustrated by case histories in our files. I would like, however, to cite only the following two:

The first was a 62 year old patient with cerebral sclerosis. He had the "flu" and in order to prevent a recurrence of his old tuberculosis he was treated with large doses of streptomycin.

A mental deterioration and a mild hemiplegia, which ceased after some time, were noticed as a possible reaction to the drug. A second transient "stroke" on the same side was seen during one of the following nights while the patient was in complete rest. Two months afterwards a third "transient hemiplegia" occurred when the man became very emotional while watching an international football match on television. Here the interesting fact was that the same "ischemic" accident occurred during rest as well as during a state of emotional upset.

A last case history may illustrate the case of a dumping syndrome in conjunction with hypertension leading to transistory hemiparetic phenomena.

A 60 year old doctor, who had had a Billroth II gastric resection performed, suffered some 40 years ago, during certain periods, from a feeling of fullness and nausea even after only a few bites of food. These complaints increased up to the time prior to his last admission. Gastric hemorrhages were frequent, with a loss of hemoglobin, yet without radiological evidence of a peptic ulcer near the stump or in the jejunum. Hypertension with blood pressure readings up to 225/115 was present. There was a transient light hemiparesis which later ceased almost completely. It was interesting to see that mild transient recurrences usually occurred during severe attacks of his dumping syndrome; these motor t.i.a. were accompanied by paraesthesias and sometimes even mild aphasia. During the dumping syndrome a drop in blood pressure and blood sugar was noted. Yet, it always remained a fairly circumscribed, and "local" reaction, combining deficits of hemodynamics and metabolism.

These case histories seem to me to underline sufficiently our concept of "local" and "relative" hypotension and the importance of the co-operation of several factors in the pathogenesis of cerebrovascular insufficiency.

References

BACHMANN, K., THEBIS, J., HEYNEN, H. P., GRAF, N.: Die drahtlose Blutdruckregistrierung unter physiologischen Bedingungen. Verh. dtsch. Ges. Kreisl.-Forsch. 33 (1967).

BERNSMEIER, A.: Probleme der Hirndurchblutung. Z. Kreisl.-Forsch. 48, 278—323 (1959).

— Komplikationen des Hirnkreislaufs bei der essentiellen Hypertonie. Verh. dtsch. Ges. Kreisl.-Forsch. 28, 137—153 (1966).

— Klinik und Therapie der Hypertonie. Münch. med. Wschr. 109, 1837—1845 (1967).

BODECHTEL, G.: Differentialdiagnose neurologischer Krankheitsbilder. Stuttgart: Thieme 1958.

CORDAY, E., ROTHENBERG, S. F.: The clinical aspects of cerebral vascular insufficiency. Ann. intern. Med. 47, 626 (1957).

Finnerty, F. A., Witkin, D., Facekas, F. J.: Cerebral hemodynamics during cerebral ischemia induced by acute hypotension. J. clin. Invest. **33**, 1277 (1954).

Hampton, J., Little, M.: Results of questionaire concerning controlled hypertension in anaesthesia. Lancet **1953**, 1299—1300.

Kety, S. S., Schmidt, C. F.: The nitrous oxide method for the quantitative determination of cerebral blood flow in man: theory, procedure and normal values. J. clin. Invest. **27**, 476—483 (1948).

Loman, J., Dameshek, W., Myerson, A., Goldman, D.: Effect of alterations in posture on the intra-arterial blood pressure in man. Arch. Neurol. Psychiat. (Chic.) **35**, 1225—1232 (1936).

Mechelke, K.: Die Labilität der Blutdruckregulierung bei nervösen Kreislaufregulations-störungen als Ausdruck der vegetativen Gesamtverfassung. Z. Psychother. u. med. Physiol. **7**, 79—83 (1957).

Menzel, W.: Der 24-Stundenrhythmus des menschlichen Blutkreislaufes. Ergebn. inn. Med. Kinderheilk. **61**, 1 (1941).

Meythaler, F.: Das Hypoglykämie-Syndrom. Medizin und Ernährung **8**, 195—201 (1967).

Müller, C.: Die Messung des Blutdrucks am Schlafenden als klinische Methode. Act. med. scand. **60**, 381—485 (1921).

Opitz, E., Schneider, M.: Über die Sauerstoffversorgung des Gehirns und den Mechanismus von Mangelwirkungen. Ergebn. Physiol. **46**, 126—260 (1950).

Pickering, G. W.: High blood pressure. London: J. A. Churchill Ltd. 1955.

Pierach, A.: Diskuss.-Bemerkungen über Siph. carot. Verh. dtsch. Ges. Kreisl.-Forsch. **19**, 225—226 (1953).

— Die lokale Hypotonie. Münch. med. Wschr. **101**, 737—739 (1959).

— Die Gefahren des Blutdruckabfalls für die sklerotische Strombahn. II. Internat. Gespräch über Angiologie 1962. Kreislaufbücherei, Band 21. Darmstadt: Steinkopff 1963.

— Über Blutdruck und Blutdruckschwankungen. Münch. med. Wschr. **105**, 873—881 (1963).

— Diskussionsbemerkung zur intraarteriellen Blutdruckmessung. Verh. dtsch. Ges. Kreisl.-Forsch. **33** (1967).

— Heynemann, K.: Der niedrige Blutdruck und die Hypotonie. Stuttgart: Enke 1959.

— — Allgemeine und lokale Hypotonie. In: Handbuch der praktischen Geriatrie, Band 1. Stuttgart: Enke 1965.

Prineas, J., Marshall, J.: Brit. med. J. **1966 I**, 14.

Richardson, D. W., Honour, A. J., Fenton, G. W., Stott, F. H., Pickering, G. W.: Variation in arterial pressure throughout the day and night. Clin. Sci. **26**, 445—460 (1964).

Schneider, M.: Über die Wiederbelebung nach Kreislaufunterbrechung. Thoraxchirurgie **6**, 2 (1958).

— Die Funktion des Herz-Kreislaufsystems. Mkurse ärztl. Fortbild. **1958**, 8.

Spatz, H.: Diskussionsbemerkung zum Siphon caroticus. Verh. dtsch. Ges. Kreisl.-Forsch. **19**, 224—225 (1953).

Wold, A.: Temporäre Rindenblindheit nach Blutungsschock. T. norske Laegeforen **2**, 184 (1964), zit. nach Med. Welt, Band XXVIII.

Zülch, K. J.: Gedanken über die Entstehung der Hirninfarkte und anderer Durchblutungs-störungen. In: Der Hirnkreislauf in Forschung und Praxis, I. Internat. Salzburg Conference 1962.

— Neuere Anschauungen über die Entstehung der cerebralen Insulte. Acta Med. Belg., Brüssel 1962, S. 890—904 (Extrait du Livre du Dr. Ludo van Bogaert).

— Zur Pathogenese des cerebrovasculären Insultes. Internist **4**, 64—70 (1963).

— Neues vom zerebralen Insult. Ärztl. Prax. **18**, 2401—2402 (1966).

— Hossmann, V.: Über die 24-Stunden-Rhythmik des menschlichen Blutdrucks. Dtsch. med. Wschr. **92**, 567—572 (1967).

— Kleihues, P., Gabe, D.: Die aktuelle Problematik auf dem Gebiet der Pathogenese, Klinik und Therapie der Hirndurchblutungsstörungen. Kongressband der II. Internat. Salzburg. Conference, S. 339—367. Wien: Verlag Brüder Hollinek 1964.

Circadian Changes of Blood Pressure and Stroke

V. Hossmann*

A prominent hypothesis of the pathogenesis of stroke due to cerebro-vascular insufficiency considers critical "hypotension" as one of the major causes (Denny-Brown; Riishede; Dalsgaard-Nielsen; Zülch). There may be, however, other factors of importance, which are correlated to this drop in blood pressure or which may even induce the hypotension (Zülch). This has not been sufficiently investigated.

It is well known that certain types of stroke start at night, mainly after midnight, similar to the coronary insufficiency or infarct. There also have been some strokes which started early in the afternoon. Both types of stroke seemed to be connected with the lows of the "biotonus" (at 3 a.m. and 2 p.m.) as expressed, for instance, by "attention" (see the curves of Graf, 1954 which were based on clerical errors from readings in a Scandinavian gas-works, see Zülch, 1962, Fig. 4). The problem then arose whether these time correlations between the beginning of "stroke" and biological "lows" were in a causal correlation.

Biological rhythms of numerous body functions have been described in the past decades (Menzel; Sollberger; Kroetz; Halberg a. m. o.). Most of them are genetically determined, but may be phase-shifted and even phase-modulated by rhythmic exogenous and/or environmental factors, such as changes of light (day-night), of noise and silence, of gravitation etc. Rhythms of 24 h duration have been called "circadian" rhythms (Halberg) and have been found in at least 40 different body functions. Moreover, some evidence exists that they are controlled by a single center, the so-called "inner clock" of Bünning. The exact localization of this center remains to be clarified.

Our long-time measurements of the blood-pressure, pulse and body temperature of a 10 year old girl, who was decerebrated due to an 8 min cardiac arrest which had occurred 7 years previously, proved that because of the well-established rhythmicity of these body functions the "inner clock" could not be localized in the neocortex (Zülch and Hossmann, 1967). Rhythmic changes in the blood pressure have been observed for a long time (Müller and Brown; Katsch and Pansdorf; Kroetz; Menzel). Rhythmic "lows" after midnight are independent of sleep and have been found also in persons accustomed to working at night, for instance in shift-work (Menzel). These "lows" are, however, abolished in malignant hypertension (Franke; Shaw et al.; Richardson et al.; Bock and Kreuzenbeck). A second low in the early afternoon has been described by Menzel (1955) in 20 of 88 patients suffering from chronic insomnia. J. Aschoff was able to show an increased circulatory lability at 3 a. m. and at midday (see Kirchhoff, 1969).

* Max-Planck-Institut für Hirnforschung, Department of General Neurology and the Neurological Department of the City Hospital Köln-Merheim.

Our Investigations

As a first step towards the clarification of whether circadian rhythms play a causative
role in stroke, blood pressure measurements were taken in our ward on "normal"
patients without apparent vascular disease. These measurements were performed
hourly or even every 30 min over at least a 24 h period. In 120 subjects a drop of
blood pressure with a deep "low" at 3 a.m. and a second less marked depression
(Fig. 1) in the early afternoon (around 2 p.m.) could be observed (for details see
Zülch and Hossmann, 1967).

After this type of oscillation had been proved, we were interested in learning
whether the same or even more marked rhythmic changes occurred in patients who
had recently suffered from cerebrovascular disorders. As with the normal patients

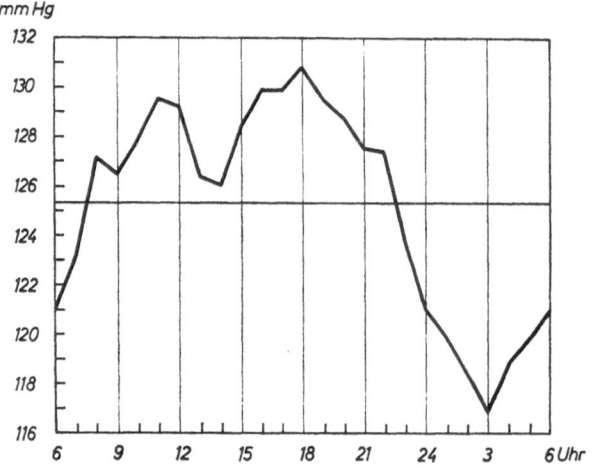

Fig. 1. Summary graph of 120 patients with normal circulation (systolic pressure)

hourly blood pressure readings were made for 24 h on 160 patients following their
admission to the hospital. The Riva-Rocci procedure was used. Automatic measurements on fat and restless patients were less reliable, an arterial puncture being not
feasible in our group of stroke patients.

The mean arterial pressure of this group was then compared with that of the
"normals". The shapes of the curves were similar (see Fig. 1 and 2), peaks at 11 a.m.
and 6 p.m. and lows at 3 a.m. and 2 p.m. being visible in both. Nevertheless, it is
remarkable to see that the low blood pressure of the stroke patients at night reaches
the medium blood pressure level only at 11 a.m. while it returns to this level by 8 a.m.
in the "normal" patients. In 17 of the 160 cases, "stroke" started in the morning upon
getting up. The "hypotonic" blood pressure in the morning in connection with
orthostatic hypotension seems to play a significant role in this type of stroke patient.
Similar observations have been made by J. S. Meyer, Leidermann and Denny-
Brown (1956). Of our stroke patients, 71% had a normotonic, 15% a hypertonic
and 14% a hypotonic arterial blood pressure level.

In 127 patients of this series precise data about the time of onset could be obtained.
In 40, i.e. about one third of this group, cerebrovascular disorders began during the

night, mainly between 1 and 5 a.m. This parallels the "low" of the mean arterial pressure and suggests that such "lows" could be a relevant cause in the pathophysiological mechanism of cerebrovascular disorders or infarct. This concept, however, could be true only if one assumed the second of the three main hypotheses of the cerebrovascular insufficiency (c.v.i.) now under discussion:

1. An angiospastic mechanism occurring at the peak of an abrupt blood pressure elevation (v. BERGMANN; FR. KAUFMANN; ZÜLCH, 1957; MARSHALL). The frequency, according to our own observations, is low (3 to 5%, ZÜLCH, 1962). It was DENNY-BROWN who rightly doubted the prevalent significancy of this mechanism.

2. Critical fall of blood pressure in patients with stenosing disease of extra- or intracranial brain arteries (DENNY-BROWN; RIISHEDE; DALSGAARD-NIELSEN; VAN DER EECKEN and ADAMS; BODECHTEL; ZÜLCH; BERNSMEYER). Failure of an auxiliary

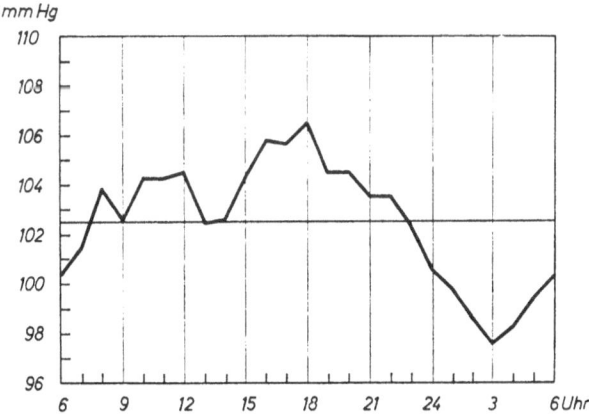

Fig. 2. Summary graph of 120 patients with normal circulation (mean arterial pressure)

collateral circulation to counteract the insufficiency of the latter (SORGO; ZÜLCH; VAN DER EECKEN and ADAMS) in case of old occlusion may be of additional importance or act as a similar mechanism.

The consequences of this hemodynamic c.v.i. for the type of infarct resulting has been worked out in detail at our institute (ZÜLCH, 1955—1969; ZÜLCH and KLEIHUES; ZÜLCH, KLEIHUES and GABE) and been confirmed by ROMANUL et al. and many others (importance of the theory of the last field (M. SCHNEIDER) or the "watershed" and "frontier zones"). It will not be discussed here in detail whether or not the fall in blood pressure must necessarily induce c.v.i. (MANGOLD et al., 1955; NOELL, 1944) due to the mechanism of "autoregulation" of the local circulation. However, in marked arteriosclerosis of general or disseminated localization this mechanism of reduction of local flow may be readily induced. Again it cannot be fully discussed here what local changes of blood flow occur as shown by the modern measurement of circulation (LASSEN; INGVAR; HØDT-RASMUSSEN et al.). ZÜLCH has discussed the pathogenesis of the hemodynamic type of c.v.i. and has come to the conclusion that a "one factor mechanism" is not very likely though it seems that in the rather complicated action hypotension may be one of the triggering factors (1967, Fig. 1).

3. Bartsch and Swank showed the possible importance of viscosity and aggregation changes during hypotension. The close correlation of hypotension and hypoglycemia is known.

We would like to discuss more thoroughly here the mechanisms which possibly act in our patients during these blood pressure oscillation.

It is interesting to note that during night the arterial oxygen pressure does not decrease although the respiratory frequency and the vital capacity of the lungs are reduced.

Mangold et al. (1955) were able to show that the reduced vital capacity is related to an increased blood deposition and a slower blood flow in the lungs. However, due to the slower blood flow in the lungs, the blood corpuscles remain in the lung capillaries longer, thus allowing 100% oxygenation even at low alveolar oxygen pressure.

The physiological blood pressure fall during night is modulated by numerous environmental factors. There are two which may play a significant role: Franke has observed that the speed of the blood pressure fall (the absolute value of which may remain constant) is dependent on the time of sleep. When sleep starts before midnight, the blood pressure fall is slow, but it will be very rapid when sleep is delayed to after midnight. The sudden drop may prevent the collateral anastomoses from efficiently working and this may reduce the collateral blood supply to the territory of the stenosed vessel. A second factor to be considered is a sleep dependent blood pressure fall which can be superimposed on the sleep independent physiological fall after midnight.- Snyder et al. showed that during REM (rapid eye movement)-sleep the blood pressure decreases.

All of these considerations may be relevant in those cases in which a midnight stroke was preceded by prodromal symptoms on previous nights. In 6 of the 40 patients which suffered stroke after midnight, such a correlation was evident. The following case report is typical for this group:

A 47 year old male (Dr. H. W.) felt exhausted during the last weeks. Upon awaking one morning, he suffered a slight stiffness of the right leg which disappeared within 15 min. The following night he awoke at 1.30 am. with symptoms of vertigo, and shortly thereafter the physician made the diagnosis of right-sided hemiparesis including the face. EEG showed a left-sided temporo-parietal delta-focus. By carotid arteriography, a complete occlusion of the m. c. a. was diagnosed. The occlusion occurred $1/2$ inch distally from its origin. The later filling showed excellent collateral blood supply by meningeal anastomoses from the anterior and posterior cerebral arteries, indicating that the thrombotic occlusion was pre-existing over a longer period. The 24 h-blood-pressure-curve that was written one day after stroke had its low at 1 to 2 am. which is exactly at the same time the first neurological symptoms appeared the day before.

In our opinion, these observations are startling enough to consider the physiological blood pressure fall during the night as a possible pathogenetic factor in this group of stroke patients. Efforts are now under way to prevent blood pressure changes in infarct endangered patients. A longtime observation will be necessary in order to ascertain the preventive value of the treatment in reducing the risk of a hemodynamic stroke. (For details see: V. Hossmann. Doktor-Dissertation der Universität zu Köln, 1969.)

References

v. Bergmann, W.: Funktionelle Pathologie. Berlin: Springer 1932.
Bernsmeier, A.: Probleme der Hirndurchblutung. Z. Kreisl.-Forsch. 48, 278—323 (1959).

Bock, K. D., Kreuzenbeck, W.: Spontaneous blood-pressure variations in hypertension; the effect of antihypertensive therapy and correlations with the incidence of complications. Antihypertensive Therapy. An Internat. Symp. Berlin-Heidelberg-New York: Springer 1966.
— — Über die Tageschwankungen des arteriellen Blutdrucks. Hochdruckforschung. Fortschr. auf dem Gebiet der Inn. Med. II. Symp. in Freiburg am 18. und 19. Juli 1964. Stuttgart: Thieme.
Bodechtel, G.: Zur Klinik der zerebralen Kreislaufstörungen (mit besonderer Berücksichtigung ihrer cardialen Genese). Verh. dtsch. Ges. Kreisl.-Forsch. 19, 109—130 (1953).
Bünning, E.: Die physiologische Uhr. Berlin-Göttingen-Heidelberg: Springer 1958.
Dalsgaard-Nielsen, T.: Survey of 1000 cases of apoplexia cerebri. Acta psychiat. scand. 30, 169—185 (1955).
Denny-Brown, D.: The treatment of recurrent cerebrovascular symptoms and the question of "vasospasm". Med. Clin. N. Amer. 35, 1457—1474 (1951).
— Meyer, J. S.: The cerebral collateral circulation. 2. Production of cerebral infarction by ischemic anoxia and its reversibility in early stages. Neurology (Minneap.) 7, 567—579 (1957).
Bartsch, W., Swank, R. L.: Der Effekt von Herzleistung und Blutdruck auf die Hämodynamik der spinalen Durchblutung. Verh. dtsch. Ges. inn. Med. 72, 1105—1110 (1967).
Franke, H.: Über den 24-Stunden-Rhythmus des Kreislaufs bei Gesunden und Herzkranken. Ber. phys.-med. Ges. Würzburg, N. F. 65, 48 (1951).
Graf, O.: Physiologische Leistungsbereitschaft und nervöse Belastung. Jb. 1954 der Max-Planck-Ges. zur Förderung der Wissenschaften e. v. 1954, 97—123.
Halberg, F., Good, R. A. Levine, H.,: Some aspects of the cardiovascular and renal circadian systems. Circulation 34, 715—717 (1966).
Høedt-Rasmussen, K., Skinhoj, E., Ewald, J., Bjerrum, J. K., Fahrenkrug, A., Lassen, N. A.: Regional cerebral blood flow in acute apoplexy. Arch. Neurol. Psychiat. (Chic.) 17, 271—281 (1967).
Ingvar, D. H.: The pathophysiology of the stroke related to findings in EEG and to measurements of regionnal cerebral blood flow. Thule Intern. Symposia, Stroke 1966, p. 104.
— Regional cerebral blood flow in cerebro-vascular disorders. Progr. Brain Res. 30, 57 (1968).
Katsch, G., Pansdorf, H.: Die Schlafbewegungen des Blutdrucks. Münch. med. Wschr. 50, 1715 (1922).
Kauffmann, Fr.: Klinisch-experimentelle Untersuchungen zum Krankheitsbild der arteriellen Hypertension. Z. exp. Med. 42, 473—495; 43, 141—169 (1925).
Kirchhoff, H. W.: Die orthostatische Kreislaufregulationsstörung. Conscientia diagnostica, 11. Konstanz: Byk-Gulden-Lomberg 1969.
Kroetz, Ch.: Ein biologischer 24-Stunden-Rhythmus des Blutkreislaufs bei Gesundheit und bei Herzschwäche, zugleich ein Beitrag zur tageszeitlichen Häufung einer akuten Kreislaufstörung. Münch. med. Wschr. 87, 284, 314 (1940).
— Die Kreislaufgröße in Gesundheit und Krankheit. Klin. Wschr. 12, 26 (1950).
Lassen, N. A.: The luxury-perfusion syndrome and its possible relation to acute metabolic acidosis localized within the brain. Lancet 2, 1113—1115 (1969).
Marshall, J.: Effects of anoxia on brain function and behaviour. Int. J. Neurol. (Montevideo) 3, 496—503 (1962).
— Le rôle des variations fonctionelles de la circulation dans les accidents cérébraux ischémiques. Symp. Internat. sur la circulation cérébrale. Paris, October 1965. Paris: Sandoz Editions 1966.
— The management of cerebrovascular disease. London: Churchill 1965.
Menzel, W.: Der 24-Stunden-Rhythmus des menschlichen Blutkreislaufes. Ergebn. inn. Med. Kinderheilk. 61, 1 (1941).
— Menschliche Tag-Nacht-Rhythmik und Schichtarbeit. Stuttgart: Bruno Schwabe und Co. Verlag 1962.
Meyer, J. S., Heidermann, H., Denny-Brown, D.: Electroencephalographic study of insufficiency of the basilar and carotid arteries in man. Neurology (Minneap.) 6, 455 —477 (1956).

Mueller, S. C., Brown, J. E.: Hourly rhythms in blood pressure in persons with normal and elevated pressures. Ann. intern. Med. **3**, 1190 (1930).

Noell, W.: Über die Durchblutung und die Sauerstoffversorgung des Gehirns. Pflügers Arch. ges. Physiol. **247**, 528—552 (1944).

Richardson, D. W., Honour, A. J., Fenton, G. W., Stott, F. H., Pickering, G. W.: Variation in arterial pressure throughout the day and night. Clin. Sci. **26**, 445—460 (1964).

Riishede, J.: Cerebral apoplexy. Acta psychiat. scand. Suppl. 118, 32 (1957).

Romanul, F. C. A., Abramowicz, A.: Changes in brain and pial vessels in arterial border zones. Arch. Neurol. Psychiat. (Chic.) **11**, 40—65 (1964).

Schneider, M.: Durchblutung und Sauerstoffversorgung des Gehirns. Verh. dtsch. Ges. Kreisl.-Forsch. **19**, 3—25 (1953).

— Zur Pathophysiologie der Hirndurchblutung. Das ärztl. Gespräch (Troponwerke) **2**, 1—20 (1965).

Shaw, D. B., Knapp, M. S., Davies, D. H.: Variations of blood-pressure in hypertensives during sleep. Lancet **1963 I**, 797—799.

Snyder, F., Hobson, J. A., Morrison, D. F., Goldfrank, F.: J. appl. Physiol. **19**, 417 —422 (1964).

Sollberger, A.: Biological rhythm research. Amsterdam-London-New York: Elsevier Publishing Co. 1965.

Sorgo, W.: Über den durch Gefäßprozeß bedingten Verschluß der A. carotis interna. Zbl. Neurochir. **4**, 161—179 (1939).

Van der Eecken, H. M., Adams, R. D.: The anatomy and functional significance of the meningeal anastomoses of the human brain. J. Neuropath. exp. Neurol. **12**, 132—157 (1953).

Zülch, K. J.: Der cerebrale Insult. In: Die essentielle Hypertonie als Schädigungsfolge. Herbsttagung des ärztl. Sachverständigenbeirats für Fragen d. Kriegsopferversorgung, — Bonn 1957.

— Gibt es Spasmen der Hirngefäße? Medizinische **14**, 622—626 (1959).

— The pathogenesis of disturbances of cerebral blood flow. Int. J. Neurol. (Montevideo) **3**, 464—482 (1962).

— Neuere Anschauungen über die Entstehung der cerebralen Insulte. Acta Med. Belg., Brüssel, 1962, 890—904 (Extrait du Livre du Dr. Ludo van Bogaert).

— Zur Pathogenese des cerebrovasculären Insultes. Internist **4**, 64—70 (1963).

— Morphology and pathogenesis of cerebral infarction. Anales del XII Congreso Latino-americano de Neurocirurgia. Symp. Internacional de Investigaciones Neurologicas, pp. 255—265. Lima 1967.

— Reconsiderations of the Clinical Problem of Cerebrovascular Insufficiency. III. Int. Salzburg Conference 1966. Springfield (Ill.): Charles C. Thomas Publ. 1970.

— Hossmann, V.: Über die 24-Stunden-Rhythmik des menschlichen Blutdrucks. Dtsch. med. Wschr. **92**, 567—572 (1967).

— — 24-hour-rhythm of human blood pressure. Germ. med. Mth. **XII**, 513—518 (1967).

— Kleihues, P.: Neuropathology of cerebral infarction. Thule Intern. Symposia, Stroke, p. 57—75. Stockholm: Nordiska Bokhandelns Förlag 1967.

— — Gabe, D.: Die aktuelle Problematik auf dem Gebiet der Pathogenese, Klinik und Therapie der Hirndurchblutungsstörungen. II. Int. Salzburg Conference 1964, S. 339 bis 367. Wien: Verlag Brüder Hollinek 1964. — Wien. med. Wschr. **116**, 494—503 (1966).

Therapy

The Possibilities of Surgical Therapy
in Cerebrovascular Disease

E. J. WYLIE*

The use of surgical therapy in cerebrovascular disease has been a matter of increasing interest and scrutiny in major medical centers throughout the world. Interest has been accelerated by the finding that the majority of patients with symptoms of cerebrovascular insufficiency have arteriosclerotic obstructive lesions in one or more of the surgically accessible extra-cranial arteries which contribute to cerebral blood flow.

During the past 13 years at the University of California, the Division of Vascular Surgery has engaged in an intensive prospective study to evaluate the results of vascular reconstructive operations in patients with cerebro-vascular insufficiency. Studies, including cerebral arteriography, were performed on more than 900 patients with various manifestations of cerebral ischemia. Over 600 of these have had revascularization operations. Follow-up examinations were carried out at 6 month intervals on both the operated and non-operated patients.

At the beginning of the study there was almost a total lack of criteria for the proper selection of surgical candidates. In many instances, the only indication for operation was the presence of an operable lesion. As experience accumulated, the risks and results of operation became more apparent. Some of the lesions originally operated upon were found to be relatively benign. Certain forms of cerebrovascular insufficiency, i.e. acute stroke with infarction, presented a risk from operation that outweighed the potential benefit. During this same period the introduction of numerous modifications in anaesthesia and surgical technic steadily decreased the incidence of operative complications.

Limitations of space do not permit a statistical tabulation of the experiences which have evolved in the current methods of patient selection and surgical technic. Accordingly it is my purpose to discuss the current concepts at the University of California that have evolved from this study.

These concepts may be considered from two points of view: the neurologic syndrome, and the location and nature of the arterial lesions, and evaluation of the results in terms of the two objectives of operation: 1. the prevention of stroke or visual loss, and 2. the relief of existing disability. We have found it useful to divide the neurologic syndromes into the following five categories.

* From the Department of Surgery, University of California School of Medicine, San Francisco.

1. *Transient ischemic attacks.* Patients in this category have had one or more episodes of temporary neurologic dysfunction. Complete recovery usually occurs within a few minutes and always by the end of 8 to 12 h.

The various cerebral and ocular manifestations of transient ischemia are well known. Although recurrence of attacks may produce apprehension and emotional disability, the primary objective of operation is the prevention of a future disabling or fatal stroke. Acheson has reported that of 82 untreated patients with transient attacks, 42 developed a stroke within 3 years (11 disabling and 7 fatal). Of 160 similar patients observed by SIEKERT at the Mayo Clinic, 33 had developed a stroke by 4 years (6 disabling and 25 fatal).

2. *Acute stroke.* Patients in this group develop a sudden neurologic deficit, usually including hemiparesis, which fails to resolve within the first 12 h. Because the most frequent causal lesion has been complete occlusion of the surgically accessible internal carotid, there originally was much enthusiasm for emergency operation to remove the occluding lesion. It is rare that such an operation is technically feasible after the first 48 h. For reasons to be described later it is our conviction that a revascularization operation for patients with an acute stroke is contraindicated until at least 30 days have passed.

3. *Chronic stroke.* These patients have survived the acute episode and remain with a permanent neurologic deficit. It is rare that the primary lesion responsible for the stroke can be removed, and even when it can, little or no improvement results. However, certain observations suggest that an occasional patient should be considered for operation. It is recognized that approximately 20% of patients with a disabling stroke will eventually have a second stroke. The second stroke, in our experience, results from occlusion of another arterial segment. We recommend therefore that cerebral arteriography be performed after 30 days have passed with particular emphasis upon demonstrating residual pre-occlusive lesions at the common carotid bifurcation. It has occasionally been observed that measurable improvement in neurologic function, vision, or mentation has resulted from operations which increase the collateral supply to the zone peripheral to the original area of infarction.

4. *Chronic low cerebral perfusion syndromes.* This represents a distinct category seldom mentioned in the neurologic literature. Many patients with single or multiple arterial obstructive lesions will have partially disabling symptoms of decreased brain function in the absence of a demonstrable traditional neurologic deficit. Common symptoms are vertigo or light headedness, lessened visual acuity, and decrease in memory and mental faculties. Lightheadedness may be episodic, appearing only with sudden changes in posture. Stenotic or occluding lesions are frequently demonstrated in one or more of the major branches from the aortic arch or at the orifices of one or both internal carotid arteries. The net effect is reduction of total cerebral blood flow. Dramatic improvement in symptoms is the usual result of removing the obstructive lesions.

5. *Asymptomatic arterial obstructive lesion.* This category is represented by the patient who has a focal bruit in one of the extra-cranial arteries, and on subsequent arteriography is found to have a local zone of advanced stenosis. There are insufficient data to support a recommendation for routine operations on such lesions. Except for the rare case of peripheral embolization, stroke from lesions in the innominate and subclavian arteries is virtually unknown. Eventual stroke from asymptomatic lesions

at the common carotid bifurcation, however, is probably more common. At this time one can only speculate whether the lowering of operative morbidity and mortality will eventually lead to the proposal that significant stenosis at the internal carotid orifice in an symptomatic patient is an indication for operation.

Surgical Considerations Based Upon the Location of the Arterial Lesion.

Aortic arch branches. The indication for operation is now limited to chronic low cerebral perfusion syndromes in which disability is prominent. The great majority of lesions produce no symptoms even when a "subclavian steal" is demonstrated radiologically. If low perfusion symptoms are associated with both proximal subclavian and internal carotid lesions, the carotid lesions are removed first. We have usually found that this in itself gives adequate relief of symptoms.

Endarterectomy has become our preference whenever a direct approach to the lesions is to be performed. Access to the innominate artery is obtained through a median sternotomy incision, and the left subclavian artery is reached by a left lateral thoracotomy incision. The origin of the right subclavian artery can generally be reached through a cervical incision without opening the thorax.

In recent years we have preferred indirect, by-passing, extra-thoracic operations whenever possible. Whenever one of the three aortic branches has a normal lumen, the cervical portion of this artery can be used as a point of origin of a by-pass graft. The saphenous vein as an arterial graft is now preferred to synthetic grafts. Hence, proximal stenosis of the innominate artery can be relieved by a saphenous vein graft brought anteriorly across the neck to the right common carotid artery from either the left common carotid or subclavian arteries.

Vertebral arteries. The operable arteriosclerotic obstructive lesions in the vertebral arteries are those which cause incomplete occlusion at the origin of the artery. When occlusion becomes complete, distal thrombus propagation extends into the relatively inaccessible portions of the artery. Increasing experience with four-vessel arteriography has narrowed the indication for operation on even the accessible proximal portion. Arteriosclerotic stenosis or total occlusion of one vertebral artery alone almost never produces symptoms of vertebro-basilar insufficiency. Bilateral stenosis also is frequently well tolerated. When vertebro-basilar symptoms do occur in the presence of vertebral lesions, there usually are additional obstructive lesions in either the basilar or carotid arteries. Since the carotid blood flow is an important source of collateral supply to the basilar artery, it is not surprising that the removal of obstructive carotid lesions will usually relieve symptoms produced by coexisting vertebral lesions.

There is little evidence to support the value of prophylactic operations for vertebral stenosis in preventing the ultimate development of a crippling or fatal neurologic deficit. We have encountered no instance in which the progression of a vertebral arteriosclerotic lesion to complete occlusion has resulted in the development of a stroke, when no other lesions are present.

The practical result of these observations has been the restriction of the surgical indications of vertebral disease to those rare instances where vertebro-basilar symptoms occur in the *absence* of operable carotid or proximal artery obstruction or when symptoms continue after such associated lesions are removed. When either of these

14*

conditions is present the vertebral lesions are consistently bilateral. The fact that we have seen only seven patients with bilateral vertebral stenosis in the absence of other obstructive lesions attests to the rarity of the need for direct vertebral artery reconstruction. In each of the seven patients, reopening of only one of the vertebral arteries caused relief of symptoms.

Of the various surgical technics proposed for the relief of proximal vertebral artery stenosis, the simplest and most effective in our experience has been transsubclavian endarterectomy. Simple removal of the core of thickened intima in the vertebral orifice restores a larger than normal lumen without the need for a patch graft.

Common carotid arteries. Stenosis or complete occlusion of the common carotid arteries is rare. As with the proximal aortic arch lesions, the symptoms are associated with decrease of total cerebral perfusion. Reduction of visual acuity is often the major symptom. In at least 50% of the patients we have seen with complete common carotid occlusion, the distal branches are patent. When there is distal patency, even if limited to the external carotid artery, dramatic improvement results from restoring patency in the common carotid artery.

Loop endarterectomy alone is the simplest technic for reopening the right common carotid artery. On the left side only the cervical portion is opened by this technic. The artery is transected at the level of the clavicle. The distal end is anastomosed to a saphenous vein graft taking origin from the distal subclavian artery.

Internal carotid arteries. The most common lesions requiring surgical intervention are those at the orifice of the internal carotid arteries. Since lesions at this site with one exception can be removed with relative safety, it is particularly important to assess the significance of operation in terms of relief of disability and prevention of stroke.

Patients with acute hemiparesis caused by internal artery occlusion are the one group in which the hazard of operation appears to outweigh the potential benefit. In a series of 28 patients operated upon within 48 h of the onset of hemiparesis, a combination of endarterectomy and distal thrombectomy restored normal blood flow in all but five. Thirteen of the patients in whom operation was technically successful developed rapid progression of the neurologic deficit within 8 h after operation, ending in coma and eventual death. Autopsy showed massive hemorrhagic infarction of the brain on the side of the carotid operation in each case. This experience suggests that it is unusually hazardous to restore normal blood flow or normal pressure to an area of brain softening and infarction.

No precise information is available concerning the value of operation for carotid stenosis in preventing ultimate stroke. Published reports on the natural history of untreated cerebro-vascular insufficiency have dealt with clinical syndromes rather than specific lesions. Information that is available suggests that arteriosclerotic lesions in the cervical portion of the internal carotid artery are more ominous than any other surgically accessible lesion. The commonest cause of hemiparesis from obstructive vascular disease is occlusion of the internal carotid artery. The data quoted earlier in this paper alluded to the high frequency of eventual stroke in patients with transient ischemic attacks. Ninety per cent of patients with transient ischemic attacks studied by us had ulcerating or stenotic lesions in the proximal internal carotid arteries. It is pertinent therefore that the primary focus of attention in cerebro-vascular surgery has been directed to be internal carotid arteries.

There appear to be at least three mechanisms whereby lesions in the internal carotid artery may lead to cerebral infarction: 1. macro-embolization from ulcerated lesions, 2. sudden decrease in cerebral perfusion caused by acute systematic circulatory depression, and 3. the progression of carotid stenosis to complete occlusion. All three mechanisms have been observed in our own experience.

Most of the patients in our study who were seen with hemiparesis, or who had a history of hemiparesis caused by internal carotid lesions, described premonitory symptoms. These included transient ischemic attacks in either the hemispheric or in the vertebro-basilar territory, recurrent ipsilateral visual episodes (amaurosis fugax), and manifestations of chronic decreased cerebral perfusion. The presence of any of these symptoms, therefore, particularly when an appropriate carotid bruit is present,

Table 1. *Long term results following carotid endarterectomy in patients with transient cerebral ischemic episodes (1957—1967)*[a]
Average length of follow-up: $^1/_2$ — 3 years

	Hemispheric Attacks	Vertebro-basilar Attacks
Total number patients	139	114
Surviving operation	132	108
Number with further transient attacks	2	17
Number developing later stroke	4	4
	(1 fatal)	(all fatal)
Total late deaths	12	28
Caused by cardiovascular disease	9	14

[a] There has been a significant reduction in the neurologic morbidity of carotid endarterectomy in the three years following compilation of these data. In 253 operations a postoperative neurologic deficit appeared in only two patients, one of which was fatal.

has become an indication for arteriography, followed by operation if an operable carotid lesion is demonstrated.

Because recurrent transient ischemic attacks have been the most common symptom in our patients operated upon for carotid stenosis, we have recently reviewed the records of 253 patients who had undergone carotid endarterectomy for this symptom. Symptoms were separated into those thought to represent vertebro-basilar insufficiency and those representing hemispheric insufficiency. In general, the patients with vertebro-basilar symptoms had additional lesions in the vertebro-basilar or proximal arteries. Symptoms were commonly produced by postural changes or by transient hypotension. Severe stenosis at the carotid lesion was uniformly found. In this group, symptoms were considered to be the result of interference with collateral supply to the posterior circulation.

The patients with pure hemispheric insufficiency included those with transient episodes of weakness or paralysis of the contralateral extremity. Many had associated attacks of ipsilateral visual field defects. Ulceration of the intima, often in the absence of severe stenosis, was the characteristic finding at operation and suggests that micro-embolization was often the cause of symptoms.

The results of operation in each group in terms of relief from subsequent attacks are illustrated in Table 1. The relatively few patients who have subsequently developed a disabling stroke is in marked contrast to the previous data on untreated patients.

The most important factor influencing the decision for surgical management is the risk of operation. Stroke has been the only surgical complication and has occurred in 3.9% of the operations (Table 2). Hemiparesis, during or immediately following operation, can be attributed to one of three causes—the dislodging of embolic fragments from the atherosclerotic lesion, inadequate collateral flow during the period of operative arterial occlusion, and postoperative thrombotic occlusion of the operated segment. Care in avoiding manipulation of the artery at the site of the atherosclerotic lesion, improvements in endarterectomy technic, and the use of general anaesthesia and hypercarbia have resulted in a significantly lessened morbidity. There have been no deaths or postoperative neurologic deficits among the last 85 consecutive carotid endarterectomy operations in 1967—1968.

Table 2. *Cerebral complications of carotid endarterectomy for transient cerebral ischemic episodes (1957—1967)*

		253 Patients (%)	332 Operations (79 bilateral lesions) (%)
Fatal strokes	(6 patients)	2.5	1.8
Non-fatal strokes	(7 patients)	2.7	2.1
	Total	5.2	3.9

Endarterectomy in the carotid artery is performed through a longitudinal arteriotomy beginning in the common carotid artery and extending to, but not beyond, the end of the bulbous portion of the internal carotid artery. Simple closure, without patching, allows restoration of flow within 12 to 14 min of occlusion and produces a smooth arterial lumen of normal size.

Summary

Of the various syndromes of cerebro-vascular insufficiency, that of transient cerebral ischemia, particularly when associated with an orifice lesion in the internal carotid artery, presents the strongest indication for vascular surgery. The relief from further attacks and the probable reduced incidence of eventual stroke greatly outweigh the now negligible operative morbidity. Lesions at other surgically accessible sites, although less ominous in terms of their stroke potential, may produce chronic or intermittent symptoms as a consequence of reduced cerebral perfusion. The indications for arterial reconstruction depend upon the need for relief of disability. The removal of an obstructive lesion is contraindicated in the presence of acute infarction.

Subject Index